Student Solution

for

Principles of Biostatistics

Second Edition

Kimberlee Gauvreau
Harvard Medical School

Marcello Pagano
Harvard School of Public Health

BROOKS/COLE
CENGAGE Learning

Australia • Brazil • Japan • Korea • Mexico • Singapore • Spain • United Kingdom • United States

BROOKS/COLE
CENGAGE Learning·

**Student Solutions Manual for
Principles of Biostatistics, Second
Edition**

**Kimberlee Gauvreau and Marcello
Pagano**

Assistant Editor: Ann Day

Marketing Manager: Tom Ziolkowski

Production Coordinator: Dorothy Bell

Cover Design: Kelly Shoemaker

Print Buyer: Micky Lawler

For product information and technology assistance,
contact us at **Cengage Learning Customer & Sales
Support, 1-800-354-9706**

For permission to use material from this text or
product, submit all requests online at
www.cengage.com/permissions
Further permissions questions can be emailed to
permissionrequest@cengage.com

ISBN-13: 978-0-534-37398-6

ISBN-10: 0-534-37398-4

For more information about this or any other Brookscole
products, contact:

Brooks/Cole
20 Channel Center Street
Boston, MA 02210
USA

Printed in Canada
10 9 8 7 6

TABLE OF CONTENTS

CHAPTER 1

The exercises in Chapter 1 are designed to provoke thought about issues that arise when numbers are used to communicate ideas. These questions do not have a single correct answer; therefore, solutions are not provided.

Similarly, solutions are not provided for the first few exercises in each chapter, which are intended to review concepts and definitions.

CHAPTER 2

Exercise 7

a. The number of suicides is discrete.
b. The concentration of lead is continuous.
c. The length of time is continuous.
d. The number of previous miscarriages is discrete.

Exercise 9

The statement is not accurate. The intervals in the table are of unequal length; therefore, it does not make sense to compare the absolute frequencies within them. The interval 16–30 has length 15 minutes, for example, while the interval 11–15 is only 5 minutes long.

Exercise 11

A bar chart of the number of executions by year is shown below. There were only a few executions in the eight years immediately following the 1976 Supreme Court decision. After that, the number of executions increased, and has continued to increase over time (although not steadily; there are periodic decreases).

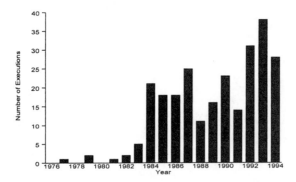

Exercise 13

a. Because the total number of smokers is not equal to the total number of nonsmokers, it is not fair to compare the distributions of absolute frequencies for these two groups.
b. The table of relative frequencies of serum cotinine levels appears below.

Cotinine Level (ng/ml)	Smokers (%)	Nonsmokers (%)
0–13	5.1	95.8
14–49	8.6	2.1
50–99	9.2	0.7
100–149	13.4	0.4
150–199	12.8	0.2
200–249	14.3	0.2
250–299	9.8	0.3
300+	26.8	0.3

c. The frequency polygons are shown below. Note that the intervals are of unequal length. For the purposes of constructing the polygons, the last interval is assumed to be 300–349 ng/ml.

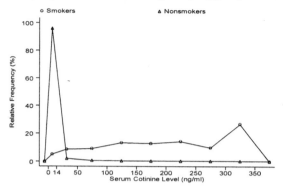

d. The distribution of smokers is fairly uniform across cotinine levels. The relative frequency is smallest in the first interval (0–13 ng/ml). It then increases, and remains consistent (hovering around 10%) across subsequent intervals up to the last (300+ ng/ml), where the relative frequency increases. For nonsmokers, nearly everyone has a cotinine level below 13 ng/ml; the relative frequency in each of the other intervals is very small.

e. Yes, it is possible that some of the subjects are misclassified. In particular, there are a number of self-reported "nonsmokers" with extremely high cotinine levels.

Exercise 15

a. The line graph is shown below.

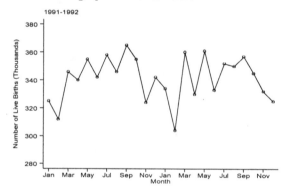

b. Based on this two-year period, there does appear to be a tendency toward more births in the spring and summer months and fewer in fall and winter.

Exercise 17

a. A box plot of the percentages of low birth weight infants is displayed on the following page.

b. The distribution is skewed to the right; its tail extends in the direction of the higher values.

c. According to the box plot, the data contain two outlying observations. In Bangladesh 50% of infants are low birth weight, and in India 33% are low birth weight.

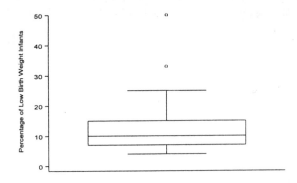

Exercise 19

a. The one-way scatter plot appears below. There are a number of instances in which two or more measurements have the same value; for example, three cigarettes have a concentration of 10 mg, five have a concentration of 13 mg, and eight have a concentration of 16 mg.

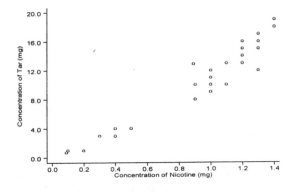

b. There are only a few values representing concentrations of tar less than 4 mg (note, however, that in two instances the line on the scatter plot actually represents two cigarettes); between 8 mg and 19 mg, the distribution is fairly uniform.

c. The two-way scatter plot is shown below.

d. There does appear to be a relationship between these quantities; the concentration of tar increases as the concentration of nicotine increases.

4

CHAPTER 3

Exercise 7

a. The mean calcium level is

$$\bar{x} = \frac{\sum_{i=1}^{8} x_i}{8}$$

$$= \frac{25.14}{8}$$

$$= 3.14 \text{ mmol/l.}$$

The median is the average of the 4th and 5th values, or

$$\frac{2.99 + 3.17}{2} = 3.08 \text{ mmol/l.}$$

The standard deviation is

$$s = \sqrt{\frac{\sum_{i=1}^{8} (x_i - \bar{x})^2}{8 - 1}}$$

$$= \sqrt{\frac{\sum_{i=1}^{8} (x_i - 3.14)^2}{7}}$$

$$= \sqrt{0.2608}$$

$$= 0.51 \text{ mmol/l.}$$

The range is the largest value minus the smallest value, or

$$3.84 - 2.37 = 1.47 \text{ mmol/l.}$$

b. The mean albumin level is 40.4 g/l. The median is 42 g/l. The standard deviation is 3.0 g/l. The range is 9 g/l.

c. The patients suffering from vitamin D intoxication all have albumin levels within the normal range. However, they do not have normal blood levels of calcium. Both the mean and the median lie above the upper limit of the normal range; overall, 6 of the 8 patients have calcium levels that are above normal.

Exercise 9

a. Europe has the smallest mean; its infant mortality rates are much lower than those of either Africa or Asia.

The box plots above the one-way scatter plots indicate that Africa has the largest median; it has the greatest proportion of nations with relatively large infant mortality rates.

Europe has the smallest standard deviation. Its infant mortality rates are more tightly clustered about the mean than those of either Africa or Asia.

b. Since the distribution of values for Africa is unimodal and roughly symmetric, the mean and the median infant mortality rates should be fairly close.

We would not expect the mean and median to be equal for Asia, however; since the distribution is skewed to the right, the mean will be larger than the median.

Exercise 11

a. The mean is 87.9 $\mu g/dl$; the median is the 50th percentile, or 86 $\mu g/dl$; the standard deviation is 16.0 $\mu g/dl$; the range is $153 - 50 = 103$ $\mu g/dl$; and the interquartile range is $98 - 76 = 22$ $\mu g/dl$.

```
. summarize zinc, detail
```

                              serum zinc level

	Percentiles	Smallest		
1%	56	50		
5%	64	51		
10%	70	53	Obs	462
25%	76	55	Sum of Wgt.	462
50%	86		Mean	87.93723
		Largest	Std. Dev.	16.00469
75%	98	142		
90%	108	147	Variance	256.1501
95%	115	151	Skewness	.6211264
99%	135	153	Kurtosis	3.890067

b. Using Chebychev's inequality, we can say that the interval $87.9 \pm (2 \times 16.0)$ or $(55.9, 119.9)$ encompasses at least 75% of the observations, the interval $87.9 \pm (3 \times 16.0)$ or $(39.9, 135.9)$ contains at least 88.9% of the observations, and the interval $87.9 \pm (4 \times 16.0)$ or $(23.9, 151.9)$ contains at least 93.8%.

c. Using the empirical rule, we would expect that 95% of the values lie within 2 standard deviations of the mean, and "almost all" lie within 3 standard deviations of the mean. In reality, $447/462 = 96.8\%$ of the observations lie in the interval $(55.9, 119.9)$, and $458/462 = 99.1\%$ in the interval $(39.9, 135.9)$.

d. Since the serum zinc levels are symmetric and unimodal (see Chapter 2, Exercise 16b), the empirical rule does a better job of summarizing the serum zinc levels than Chebychev's inequality.

Exercise 13

a. The mean concentration of nicotine is 0.99 mg and the median is 1.1 mg.

b. The histogram is shown on the following page. The distribution of values is skewed to the left.

c. Because the distribution is skewed, the median may provide a better measure of central tendency than the mean. (In this case, however, the two summary measures are very close.)

. summarize nicotine, detail

                    concentration per cig (mg)

            Percentiles   Smallest
    1%   .09   .09
    5%   .1   .1
   10%   .3   .2   Obs   35
   25%   .9   .3   Sum of Wgt.   35

   50%   1.1   Mean   .9908571
                               Largest   Std. Dev.   .3883326
   75%   1.3   1.3
   90%   1.3   1.3   Variance   .1508022
   95%   1.4   1.4   Skewness   -1.199142
   99%   1.4   1.4   Kurtosis   3.132104

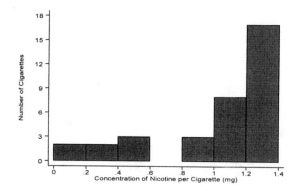

Exercise 15

a. The box plots of systolic blood pressure measurements are shown below. The two distributions of values are quite similar; in particular, the 25th, 50th, and 75th percentiles are nearly the same. The males have two high outlying values, while the females have one low outlier.

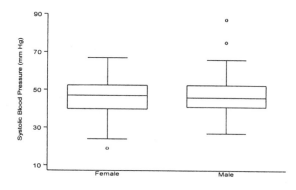

b. For girls, the mean systolic blood pressure is $\bar{x}_f = 46.5$ mm Hg and the standard deviation is $s_f = 11.1$ mm Hg; for boys, the mean is $\bar{x}_m = 47.9$ mm Hg and the standard deviation is $s_m = 11.8$ mm Hg. Males have a slightly larger mean and standard deviation.

```
. sort sex
. by sex: summarize sbp

-> sex=  Female
Variable |    Obs        Mean    Std. Dev.      Min         Max
---------+-------------------------------------------------------
     sbp |     56    46.46429    11.14526        19          67

-> sex=  Male
Variable |    Obs        Mean    Std. Dev.      Min         Max
---------+-------------------------------------------------------
     sbp |     44    47.86364    11.80577        27          87
```

c. The coefficient of variation for females is

$$CV_f = \frac{s_f}{\bar{x}_f} \times 100\%$$

$$= \frac{11.15}{46.46} \times 100\%$$

$$= 24\%,$$

and the coefficient of variation for males is

$$CV_m = \frac{s_m}{\bar{x}_m} \times 100\%$$

$$= \frac{11.81}{47.86} \times 100\%$$

$$= 25\%.$$

There is no evidence that the amount of variability in systolic blood pressure differs for males and females.

CHAPTER 4

Exercise 7
The statement is misleading. Although the number of deaths has been increasing, the population base could be increasing as well.

Exercise 9
a. The infant mortality rates for each category of birth weight appear in the table below.

Birth Weight (grams)	Infant Mortality Rate (per 1000 live births)
2500+	4.5
1500–2499	30.0
750–1499	233.1
500–749	765.7
<500	885.9
Unknown	216.1

b. Infant mortality rate increases as birth weight decreases.
c. The infant mortality rates for children weighing ≥ 1500 grams at birth — 30.0 per 1000 for those weighing 1500–2499 grams and 4.5 per 1000 for those weighing ≥ 2500 grams — are much lower than the mortality rate for those with unknown birth weight (216.1 per 1000). Consequently, a large proportion of the infants with unknown birth weight are likely to have weighed < 1500 grams.

Exercise 11
a. The rates of reported cases of polio per 100,000 children are 40.8 for the vaccine group and 80.5 for the placebo group.
The rates of true instances of polio per 100,000 children are 28.4 for the vaccine group and 70.6 for the placebo group.
The rates of incorrect diagnoses of polio per 100,000 children are 12.5 for the vaccine group and 9.9 for the placebo group.
The rates of paralytic disease per 100,000 children are 16.4 for the vaccine group and 57.1 for the placebo group.
The rates of nonparalytic disease per 100,000 children are 12.0 for the vaccine group and 13.4 for the placebo group.
b. The Salk vaccine appears to have help prevented cases of paralytic polio. It had no apparent effect on nonparalytic polio.

Exercise 13
If the age-adjusted mortality rate is decreasing faster than the crude mortality rate, this must mean that the population is growing older over time.

Exercise 15

a. The crude cancer mortality rate for 1940 is

$$\frac{158,200}{131,670,000} = 0.0012015$$

$$= 120.2 \text{ per } 100,000 \text{ population,}$$

and the crude rate for 1986 is

$$\frac{469,330}{241,097,000} = 0.0019466$$

$$= 194.7 \text{ per } 100,000 \text{ population.}$$

The crude cancer mortality rate in 1986 is considerably higher than the crude rate in 1940. In fact, the crude rate increases

$$\frac{|\,120.2 - 194.7\,|}{120.2} = 62.0\%$$

in 46 years.

b. The proportions of the total population in each age group are displayed below.

| | Proportion | |
Age	1940	1986
0– 4	8.0%	7.5%
5–14	17.0%	14.0%
15–24	18.2%	16.2%
25–34	16.2%	17.7%
35–44	13.9%	13.7%
45–54	11.8%	9.5%
55–64	8.0%	9.2%
65–74	4.8%	7.2%
75+	2.0%	4.9%
Total	100.0%	100.0%

The population in 1986 is older than the population in 1940.

c. The age-specific cancer death rates for each population are below.

| | Death Rate per 100,000 | |
Age	1940	1986
0– 4	4.7	3.7
5–14	3.0	3.4
15–24	5.4	5.4
25–34	17.3	13.1
35–44	61.1	45.3
45–54	168.8	165.7
55–64	369.6	444.4
65–74	695.1	847.0
75+	1,183.5	1,363.5

Disregarding the effect of infant mortality, cancer death rate increases with age.

d. It is necessary to control for the effect of age when comparing death rates. The 1986 population is older than the 1940 population, and death rate increases with age. As a result, the crude cancer mortality rate for 1986 appears very high. Age is a confounder in this situation; it is associated with both the population distribution and death rate.

e. To apply the direct method of standardization, we use the equation

$$\text{expected deaths} = (\text{standard population}) \times (\text{actual age-specific death rates}).$$

Age	Expected Number of Deaths	
	1940	1986
0–4	494	386.8
5–14	667	771.8
15–24	1,287	1,296.6
25–34	3,696	2,795.4
35–44	11,198	8,310.6
45–54	26,180	25,700.4
55–64	39,071	46,984.8
65–74	44,328	54,013.5
75+	31,279	36,036.7
Total	158,200	176,296.6

The age-adjusted death rates are calculated as

$$\text{age-adjusted death rate} = \frac{\text{expected deaths}}{\text{total standard population}}.$$

Therefore, the age-adjusted cancer mortality rate for 1940 is

$$\frac{158,200}{131,670,000} = 0.0012015$$

$$= 120.2 \text{ per } 100,000 \text{ population,}$$

and the age-adjusted rate for 1986 is

$$\frac{176,296.6}{131,670,000} = 0.0013389$$

$$= 133.9 \text{ per } 100,000 \text{ population.}$$

f. Because we are using the 1940 population as the standard, the crude and age-adjusted mortality rates for that year are identical.

For 1986, the age-adjusted death rate is much lower than the crude death rate. The standard 1940 population is younger than the 1986 population, and younger age groups have lower death rates. Note that the age-adjusted death rate increases only

$$\frac{|\,120.2 - 133.9\,|}{120.2} = 11.4\%$$

over the 46-year period.

g. The age-specific cancer mortality rates are plotted on the following page. In general, the age-specific death rates follow the same trends in both 1940 and 1986. Therefore, it is not inappropriate to use the direct method to standardize for age.

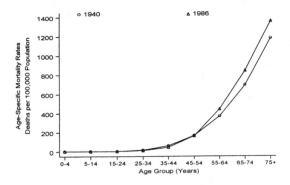

h. To apply the indirect method of standardization, we use the formula

$$\text{expected deaths} = (\text{standard age-specific death rates}) \times (\text{actual population}).$$

	Expected Number of Deaths	
Age	1940	1986
0– 4	494	850.7
5–14	667	1,006.8
15–24	1,287	2,099.3
25–34	3,696	7,409.5
35–44	11,198	20,199.5
45–54	26,180	38,505.5
55–64	39,071	82,162.9
65–74	44,328	120,478.7
75+	31,279	140,075.0
Total	158,200	412,787.9

Since the standard mortality ratio is calculated as

$$\text{standard mortality ratio} = \frac{\text{observed deaths}}{\text{expected deaths}},$$

the smr for 1940 is

$$\frac{158,200}{158,200} = 1.00$$
$$= 100\%$$

and the smr for 1986 is

$$\frac{469,330}{412,787.9} = 1.14$$
$$= 114\%.$$

i. The 1986 population has a 14% higher mortality rate than the 1940 population.

j. The age-adjusted cancer mortality rate for 1940 is

$$\frac{120.2}{100,000} \times 1.00 \;=\; 120.2 \text{ per } 100,000 \text{ population,}$$

and the age-adjusted rate for 1986 is

$$\frac{120.2}{100,000} \times 1.14 \;=\; 137.0 \text{ per } 100,000 \text{ population.}$$

The results obtained using the indirect method do correspond to those obtained when the direct method was applied; in both cases, the age-adjusted mortality rate for 1986 is somewhat higher than the age-adjusted rate for 1940. We would not expect the results to be exactly the same.

CHAPTER 5

Exercise 7
a. At the ages when males have a higher rate of accidental and violent death than females, we would see higher age-specific death rates among the males.

b. Since $\overset{\circ}{e}_x$ reflects the mortality rates for age x and beyond, we would expect the average life expectancies of males to be lower than those of females for these early age groups.

Exercise 9
Life expectancies at birth and at age 60 increased consistently over the centuries for both males and females. Life expectancies at age 80 also increased for females; for males, the life expectancy first dropped and then increased. Females have a longer life expectancy than males at each age and in every time period. The difference between genders is largest in the most recent time period, 1971–1975.

Exercise 11
a. The probability of surviving from birth to age 80 is

$$\frac{l_{80}}{l_0} = \frac{48,460}{100,000}$$
$$= 0.485.$$

b. The probability that a 50-year-old survives to age 80 is

$$\frac{l_{80}}{l_{50}} = \frac{48,460}{92,562}$$
$$= 0.524.$$

c. The probability of surviving from birth to age 10 is

$$\frac{l_{10}}{l_0} = \frac{98,877}{100,000}$$
$$= 0.989.$$

The probability of surviving from birth to age 30 is

$$\frac{l_{30}}{l_0} = \frac{97,237}{100,000}$$
$$= 0.972.$$

The probability of surviving from birth to age 50 is

$$\frac{l_{50}}{l_0} = \frac{92,562}{100,000}$$
$$= 0.926.$$

d. The probability of surviving from age 1 to age 10 is

$$\frac{l_{10}}{l_1} = \frac{98,877}{99,149}$$
$$= 0.997.$$

The probability of surviving from age 1 to age 30 is

$$\frac{l_{30}}{l_1} = \frac{97,237}{99,149}$$
$$= 0.981.$$

The probability of surviving from age 1 to age 50 is

$$\frac{l_{50}}{l_1} = \frac{92,562}{99,149}$$
$$= 0.934.$$

e. The probability that a 25-year-old will survive 10 years is

$$\frac{l_{35}}{l_{25}} = \frac{96,493}{97,825}$$
$$= 0.986.$$

The probability that a 45-year-old will survive 10 years is

$$\frac{l_{55}}{l_{45}} = \frac{89,971}{94,280}$$
$$= 0.954.$$

The probability that a 65-year-old will survive 10 years is

$$\frac{l_{75}}{l_{65}} = \frac{61,489}{80,145}$$
$$= 0.767.$$

f. The probability that a 10-year-old will survive 20 years is

$$\frac{l_{30}}{l_{10}} = \frac{97,237}{98,877}$$
$$= 0.983.$$

The probability that a 10-year-old will survive 40 years is

$$\frac{l_{50}}{l_{10}} = \frac{92,562}{98,877}$$
$$= 0.936.$$

The probability that a 10-year-old will survive 60 years is

$$\frac{l_{70}}{l_{10}} = \frac{72,063}{98,877}$$
$$= 0.729.$$

Exercise 13

a. For each age group except the first, the 1940 mortality rate in column 2 is higher than the corresponding rate in column 3; therefore, the true mortality rates in column 3 should yield the longer life expectancy at birth.

b. The life table that uses the 1940 age-specific death rates (column 2) appears below.

x to $x+n$	nq_x	l_x	nd_x	nL_x	T_x	$\overset{\circ}{e}_x$
0– 1	0.0549	100,000	5,490	97,255	6,288,414	62.9
1– 5	0.0115	94,510	1,087	375,866	6,191,159	65.5
5–10	0.0055	93,423	514	465,830	5,815,293	62.2
10–15	0.0050	92,909	465	463,383	5,349,463	57.6
15–20	0.0085	92,444	786	460,255	4,886,080	52.9
20–25	0.0119	91,658	1,091	455,563	4,425,825	48.3
25–30	0.0139	90,567	1,259	449,688	3,970,262	43.8
30–35	0.0169	89,308	1,509	442,768	3,520,574	39.4
35–40	0.0218	87,799	1,914	434,210	3,077,806	35.1
40–45	0.0301	85,885	2,585	422,963	2,643,596	30.8
45–50	0.0427	83,300	3,557	407,608	2,220,633	26.7
50–55	0.0624	79,743	4,976	386,275	1,813,025	22.7
55–60	0.0896	74,767	6,699	357,088	1,426,750	19.1
60–65	0.1270	68,068	8,645	318,728	1,069,662	15.7
65–70	0.1812	59,423	10,767	270,198	750,934	12.6
70–75	0.2704	48,656	13,157	210,388	480,736	9.9
75–80	0.3946	35,499	14,008	142,475	270,348	7.6
80–85	0.5941	21,491	12,768	75,535	127,873	6.0
85+	1.0000	8,723	8,723	52,338	52,338	6.0

The life table using the true age-specific death rates (column 3) is below.

x to $x+n$	nq_x	l_x	nd_x	nL_x	T_x	$\overset{\circ}{e}_x$
0– 1	0.0549	100,000	5,490	97,255	7,039,663	70.4
1– 5	0.0101	94,510	955	376,130	6,942,408	73.5
5–10	0.0038	93,555	356	466,885	6,566,278	70.2
10–15	0.0028	93,199	261	465,343	6,099,393	65.4
15–20	0.0056	92,938	520	463,390	5,634,050	60.6
20–25	0.0053	92,418	490	460,865	5,170,660	55.9
25–30	0.0077	91,928	708	457,870	4,709,795	51.2
30–35	0.0077	91,220	702	454,345	4,251,925	46.6
35–40	0.0123	90,518	1,113	449,808	3,797,580	42.0
40–45	0.0106	89,405	948	444,655	3,347,77	37.4
45–50	0.0247	88,457	2,185	436,823	2,903,117	32.8
50–55	0.0300	86,272	2,588	424,890	2,466,294	28.6
55–60	0.0473	83,684	3,958	408,525	2,041,404	24.4
60–65	0.0728	79,726	5,804	384,120	1,632,879	20.5
65–70	0.1055	73,922	7,799	350,113	1,248,759	16.9
70–75	0.1568	66,123	10,368	304,695	898,646	13.6
75–80	0.2288	55,755	12,757	246,883	593,951	10.6
80–85	0.3445	42,998	14,813	177,958	347,068	8.1
85+	1.0000	28,185	28,185	169,110	169,110	6.0

The difference between the two life expectancies at birth is $70.4 - 62.9 = 7.5$ years.

c. Many government and public health organizations use $\overset{\circ}{e}_x$ to help in planning for future services. Therefore, if individuals live considerably longer than anticipated based on life table calculations, these services are almost sure to be inadequate. In addition, since people will be paying premiums for several more years than calculated, insurance companies will substantially increase their profits.

Exercise 15

a. The line graph is shown below.

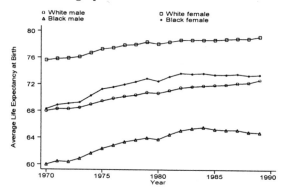

b. Females have longer life expectancies than males for both races. Whites have longer life expectancies than blacks of the same gender.

c. Over this period, life expectancy increased by 0.9 years for white males and by 0.5 years for white females. It decreased by 0.8 years for black males and by 0.2 years for black females. Among individuals in the younger age groups, mortality rates must be higher for blacks than for whites. To determine why this is the case, we could begin by examining age-specific mortality rates for specific causes of death (heart disease, AIDS, homicide, etc.) and comparing them for blacks and whites.

CHAPTER 6

Exercise 7

a. $A \cap B$ is the event that the individual is exposed to high levels of both carbon monoxide and nitrogen dioxide.

b. $A \cup B$ is the event that the individual is exposed to either carbon monoxide or nitrogen dioxide or both.

c. A^C is the event that the individual is not exposed to high levels of carbon monoxide.

d. The events A and B are not mutually exclusive.

Exercise 9

a. The probability that a woman who gave birth in 1992 was 24 years of age or younger is

$$
\begin{aligned}
P(\leq 24) &= P(< 15 \text{ or } 15\text{--}19 \text{ or } 20\text{--}24) \\
&= P(< 15) + P(15\text{--}19) + P(20\text{--}24) \\
&= 0.003 + 0.124 + 0.263 \\
&= 0.390.
\end{aligned}
$$

b. The probability that the woman was 40 years of age or older is

$$
\begin{aligned}
P(\geq 40) &= P(40\text{--}44 \text{ or } 45\text{--}49) \\
&= P(40\text{--}44) + P(45\text{--}49) \\
&= 0.014 + 0.001 \\
&= 0.015.
\end{aligned}
$$

c. Given that the woman was under 30 years of age, the probability that she was not yet 20 is

$$
\begin{aligned}
P(< 20 \,|< 30) &= \frac{P(< 20 \text{ and } < 30)}{P(< 30)} \\
&= \frac{P(< 20)}{P(< 30)} \\
&= \frac{0.003 + 0.124}{0.003 + 0.124 + 0.263 + 0.290} \\
&= 0.187.
\end{aligned}
$$

d. Given that the woman was 35 years of age or older, the probability that she was under 40 is

$$
\begin{aligned}
P(< 40 \,|\geq 35) &= \frac{P(< 40 \text{ and } \geq 35)}{P(\geq 35)} \\
&= \frac{P(35--39)}{P(\geq 35)} \\
&= \frac{0.085}{0.085 + 0.014 + 0.001} \\
&= 0.850.
\end{aligned}
$$

18

Exercise 11

a. Since these events are independent, the probability that both adults are uninsured is

$$P(\text{both uninsured}) = P(\text{woman uninsured}) \times P(\text{man uninsured})$$
$$= 0.123 \times 0.123$$
$$= 0.015.$$

b. The probability that both adults are insured is

$$P(\text{both insured}) = (1 - 0.123) \times (1 - 0.123)$$
$$= 0.877 \times 0.877$$
$$= 0.769.$$

c. The probability that all five adults are uninsured is $0.123 \times 0.123 \times 0.123 \times 0.123 \times 0.123$ $= 0.000028$.

Exercise 13

a. The probability of a false negative result is

$$P(- \text{ test} \mid \text{ca}) = 1 - \text{sensitivity}$$
$$= 1 - 0.85$$
$$= 0.15.$$

b. The probability of a false positive result is

$$P(+ \text{ test} \mid \text{no ca}) = 1 - \text{specificity}$$
$$= 1 - 0.80$$
$$= 0.20.$$

c. Since $P(\text{ca}) = 0.0025$ and $P(\text{no ca}) = 0.9975$, the probability that a woman has breast cancer given that her mammogram is positive is

$$P(\text{ca} \mid + \text{ test}) = \frac{P(\text{ca})P(+ \text{ test} \mid \text{ca})}{P(\text{ca})P(+ \text{ test} \mid \text{ca}) + P(\text{no ca})P(+ \text{ test} \mid \text{noca})}$$
$$= \frac{(0.0025)(0.85)}{(0.0025)(0.85) + (0.9975)(0.20)}$$
$$= 0.0105.$$

Exercise 15

a. The sensitivity of radionuclide ventriculography is

$$P(+\text{rv} \mid \text{cad}) = \frac{302}{481}$$
$$= 0.628,$$

and its specificity is

$$P(-\text{rv} \mid \text{no cad}) = \frac{372}{452}$$
$$= 0.823.$$

b. Since P(cad) = 0.10 and P(no cad) = 0.90, the probability that an individual has coronary artery disease given that he or she tests positive is

$$P(\text{cad} \mid +\text{rv}) = \frac{P(+\text{rv} \mid \text{cad})P(\text{cad})}{P(+\text{rv} \mid \text{cad})P(\text{cad}) + P(+\text{rv} \mid \text{no cad})P(\text{no cad})}$$

$$= \frac{(0.628)(0.10)}{(0.628)(0.10) + (0.177)(0.90)}$$

$$= 0.283.$$

c. The predictive value of a negative test is

$$P(\text{no cad} \mid -\text{rv}) = \frac{P(-\text{rv} \mid \text{no cad})P(\text{no cad})}{P(-\text{rv} \mid \text{no cad})P(\text{no cad}) + P(-\text{rv} \mid \text{cad})P(\text{cad})}$$

$$= \frac{(0.823)(0.90)}{(0.823)(0.90) + (0.372)(0.10)}$$

$$= 0.952.$$

Exercise 17

a. In Brooklyn, the probability of a positive test result is 0.0129, or 1.29%.

b. The prevalence of HIV infection is

$$P(H) = \frac{P(+\text{ test}) - [1 - \text{specificity}]}{\text{sensitivity} - [1 - \text{specificity}]}$$

$$= \frac{0.0129 - [1 - 0.998]}{0.99 - [1 - 0.998]}$$

$$= 0.011.$$

Exercise 19

a. The probabilities of suffering from persistent respiratory symptoms by socioeconomic status are shown below.

Socioeconomic Status	Probability
Low	0.392
Middle	0.238
High	0.141

b. Let S represent the presence of symptoms. The odds of experiencing persistent respiratory symptoms for the middle group relative to the high group are

$$\text{OR} = \frac{P(S \mid \text{middle})/[1 - P(S \mid \text{middle})]}{P(S \mid \text{high})/[1 - P(S \mid \text{high})]}$$

$$= \frac{(0.238)/(1 - 0.238)}{(0.141)/(1 - 0.141)}$$

$$= 1.90,$$

and for the low group relative to the high group are

$$OR = \frac{P(S \mid \text{low})/[1 - P(S \mid \text{low})]}{P(S \mid \text{high})/[1 - P(S \mid \text{high})]}$$

$$= \frac{(0.392)/(1 - 0.392)}{(0.141)/(1 - 0.141)}$$

$$= 3.93.$$

c. There does appear to be an association between socioeconomic status and respiratory symptoms; the odds of experiencing symptoms increase as socioeconomic status decreases.

CHAPTER 7

Exercise 9

It is unlikely that X has a binomial distribution. While there are a fixed number of trials ($n = 7$) that each result in one of two mutually exclusive outcomes, the outcomes of the trials are not independent. If the concentration of carbon monoxide is exceptionally high one day, the pollution will not all disappear over night; the concentration is more likely to be high the next day as well.

Binomial Dis: Fixed # of Trials = p
Trials : mutually Exclusive
Trial: Independent

Exercise 11

a. The ten persons can be ordered in $10! = 3,628,800$ different ways.

b. Four individuals can be selected in

$$\binom{10}{4} = \frac{10!}{4!\,6!}$$
$$= 210$$

different ways.

c. The probability that exactly three of the ten individuals are left-handed is

$$P(\text{three left-handers}) = \binom{10}{3}(0.098)^3(0.902)^{10-3}$$
$$= 0.055.$$

d. The probability that at least six of the ten persons are left-handed is

$$P(\text{at least six}) = P(\text{six}) + P(\text{seven}) + P(\text{eight}) + P(\text{nine}) + P(\text{ten})$$
$$= \binom{10}{6}(0.098)^6(0.902)^4 + \binom{10}{7}(0.098)^7(0.902)^3$$
$$+ \binom{10}{8}(0.098)^8(0.902)^2 + \binom{10}{9}(0.098)^9(0.902)^1$$
$$+ \binom{10}{10}(0.098)^{10}(0.902)^0$$
$$= 0.0001.$$

e. The probability that at most two individuals are left-handed is

$$P(\text{at most two}) = P(\text{none}) + P(\text{one}) + P(\text{two})$$
$$= \binom{10}{0}(0.098)^0(0.902)^{10} + \binom{10}{1}(0.098)^1(0.902)^9$$
$$+ \binom{10}{2}(0.098)^2(0.902)^8$$
$$= 0.933.$$

EX 12 → NO

Exercise 13

a. The number of children who become infected follows a binomial distribution. The mean number of infected children per sample is $np = 224(0.25) = 56$.

b. The standard deviation is $\sqrt{np(1-p)} = \sqrt{244(0.25)(0.75)} = 6.5$.

c. According to Chebychev's inequality, we would expect the interval $56 \pm 2(6.5)$ or $(43, 69)$ to contain at least 75% of the observations, and the interval $56 \pm 3(6.5)$ or $(36.5, 75.5)$ to contain at least 88.9%.

Exercise 15

a. The probability that no suicides will be reported is

$$P(\text{no suicides}) = \frac{e^{-2.75}(2.75)^0}{0!}$$
$$= 0.064.$$

b. The probability that at most four suicides will be reported is

$$
\begin{aligned}
P(\text{at most four}) &= P(\text{none}) + P(\text{one}) + P(\text{two}) + P(\text{three}) + P(\text{four}) \\
&= \frac{e^{-2.75}(2.75)^0}{0!} + \frac{e^{-2.75}(2.75)^1}{1!} + \frac{e^{-2.75}(2.75)^2}{2!} \\
&\quad + \frac{e^{-2.75}(2.75)^3}{3!} + \frac{e^{-2.75}(2.75)^4}{4!} \\
&= 0.855.
\end{aligned}
$$

c. The probability that six or more suicides will be reported is

$$
\begin{aligned}
P(\text{six or more}) &= 1 - P(\text{less than six}) \\
&= 1 - [P(\text{none}) + P(\text{one}) + P(\text{two}) + P(\text{three}) \\
&\quad + P(\text{four}) + P(\text{five})] \\
&= 1 - \left[0.855 + \frac{e^{-2.75}(2.75)^5}{5!} \right] \\
&= 1 - 0.939 \\
&= 0.061.
\end{aligned}
$$

Exercise 17

a. The probability that z is greater than 2.60 is 0.005.

b. The probability that z is less than 1.35 is $1 - 0.089 = 0.911$.

c. The probability that z is between -1.70 and 3.10 is $1 - 0.045 - 0.001 = 0.954$.

d. The value $z = 1.04$ cuts off the upper 15% (actually, 14.9%) of the standard normal distribution.

e. The value $z = -0.84$ cuts off the lower 20% of the distribution.

Exercise 19

a. The probability that a randomly selected man weighs less than 130 pounds is

$$
\begin{aligned}
P(X < 130) &= P\left(\frac{X - 172.2}{29.8} < \frac{130 - 172.2}{29:8} \right) \\
&= P(Z < -1.42) \\
&= 0.078.
\end{aligned}
$$

23

b. The probability that he weighs more than 210 pounds is

$$P(X > 210) = P\left(\frac{X - 172.2}{29.8} > \frac{210 - 172.2}{29.8}\right)$$
$$= P(Z > 1.27)$$
$$= 0.102.$$

c. Among five males selected at random, the probability that at least one will have a weight outside the range 130 to 210 pounds is

$$P(\text{at least one} < 130 \text{ or} > 210) = 1 - P(\text{none} < 130 \text{ or} > 210)$$
$$= 1 - P(\text{all between 130 and 210})$$
$$= 1 - [P(130 \leq X \leq 210)]^5$$
$$= 1 - [1 - 0.078 - 0.102]^5$$
$$= 1 - (0.820)^5$$
$$= 0.629.$$

CHAPTER 8

Exercise 9

a. The distribution of means of samples of size 10 has mean $\mu = 0$, standard error $\sigma/\sqrt{n} = 1/\sqrt{10} = 0.32$, and is normally distributed. (Since the underlying population is itself normal, this is true for any sample size n.)

b. The proportion of means that are greater than 0.60 is

$$P(\overline{X} > 0.60) = P\left(\frac{\overline{X} - 0}{0.32} > \frac{0.60 - 0}{0.32}\right)$$
$$= P(Z > 1.87)$$
$$= 0.031$$
$$= 3.1\%.$$

c. The proportion of means that are less than -0.75 is

$$P(\overline{X} < 0.75) = P\left(\frac{\overline{X} - 0}{0.32} < \frac{0.75 - 0}{0.32}\right)$$
$$= P(Z < -2.34)$$
$$= 0.010$$
$$= 1.0\%.$$

d. The value $Z = 0.84$ cuts off the upper 20% of the standard normal distribution. Therefore, $\overline{X} = 0.84(0.32) + 0 = 0.27$ cuts off the upper 20% of the distribution of sample means.

e. The value $Z = -1.28$ cuts off the lower 10% of the standard normal distribution, and $\overline{X} = -1.28(0.32) + 0 = -0.41$ cuts off the lower 10% of the distribution of sample means.

Exercise 11

a. The probability that the newborn's birth weight is less than 2500 grams is

$$P(X < 2500) = P\left(\frac{X - 3500}{430} < \frac{2500 - 3500}{430}\right)$$
$$= P(Z < -2.34)$$

what value cuts off the lower

$$= 0.010.$$

b. The value $Z = -1.645$ cuts off the lower 5% of the standard normal curve. Therefore, $X = (-1.645)(430) + 3500 = 2793$ cuts off the lower 5% of the distribution of birth weights.

c. The distribution of means of samples of size 5 had mean $\mu = 3500$ grams, standard error $\sigma/\sqrt{n} = 430/\sqrt{5} = 192$ grams, and is approximately normally distributed.

d. The value $\overline{X} = (-1.645)(192) + 3500 = 3184$ cuts off the lower 5% of the distribution of samples of size 5.

e. The probability that the sample mean is less than 2500 grams is

$$P(\overline{X} < 2500) = P\left(\frac{\overline{X} - 3500}{192} < \frac{2500 - 3500}{192}\right)$$
$$= P(Z < -5.21)$$
$$= 0.000.$$

f. The number of newborns with a birth weight less than 2500 grams follows a binomial distribution with $n = 5$ and $p = 0.01$. Therefore, the probability that only one of the 5 newborns has a birth weight less than 2500 grams is

$$P(X = 1) = \binom{5}{1}(0.01)^1(0.99)^4$$
$$= 0.048.$$

Exercise 13

a. Note that

$$P(300 \leq X \leq 400) = P\left(\frac{300 - 341}{79} \leq \frac{X - 341}{79} \leq \frac{400 - 341}{79}\right)$$
$$= P(-0.52 \leq Z \leq 0.75)$$
$$= 1 - 0.302 - 0.227$$
$$= 0.471.$$

Approximately 47.1% of the males have a serum uric acid level between 300 and 400 μmol/l.

b. The distribution of means of samples of size 5 is normal with mean $\mu = 341$ μmol/l and standard error $\sigma/\sqrt{n} = 79/\sqrt{5} = 35.3$ μmol/l. Therefore,

$$P(300 \leq \overline{X} \leq 400) = P\left(\frac{300 - 341}{35.3} \leq \frac{\overline{X} - 341}{35.3} \leq \frac{400 - 341}{35.3}\right)$$
$$= P(-1.16 \leq Z \leq 1.67) \qquad = P(Z \geq -1.16) - P(Z \geq 1.67)$$
$$= 1 - 0.123 - 0.047 \qquad = 1 - P(Z \geq 1.16) - P(Z \geq 1.67)$$
$$= 0.830. \qquad\qquad =$$

Approximately 83.0% of the samples have a mean serum uric acid level between 300 and 400 μmol/l.

c. The distribution of means of samples of size 10 is normal with mean $\mu = 341$ μmol/l and standard error $\sigma/\sqrt{n} = 79/\sqrt{10} = 25.0$ μmol/l. Therefore,

$$P(300 \leq \overline{X} \leq 400) = P\left(\frac{300 - 341}{25.0} \leq \frac{\overline{X} - 341}{25.0} \leq frac400 - 34125.0\right)$$
$$= P(-1.64 \leq Z \leq 2.36)$$
$$= 1 - 0.051 - 0.009$$
$$= 0.940.$$

Approximately 94.0% of the samples have a mean serum uric acid level between 300 and 400 μmol/l.

d. For the standard normal distribution, the interval $(-1.96, 1.96)$ contains 95% of the observations. The corresponding values of $\overline{X}$ are $\overline{X} = -1.96(25.0) + 341 = 292$ and $\overline{X} = 1.96(25.0) + 341 = 390$. Therefore, the interval $(292, 390)$ encloses 95% of the means of samples of size 10. This symmetric interval is shorter than an asymmetric one.

Exercise 15

The probability that a sample mean lies in the interval $(195.9, 226.1)$ is

$$P(195.9 \leq \overline{X} \leq 226.1) = P\left(\frac{195.9 - 211}{9.2} \leq \frac{\overline{X} - 211}{9.2} \leq \frac{226.1 - 211}{9.2}\right)$$

$$= P(-1.64 \leq Z \leq 1.64)$$

$$= 1 - 0.051 - 0.051$$

$$= 0.898.$$

CHAPTER 9

Exercise 5

a. A two-sided 95% confidence interval for μ_s is

$$-1.96 \leq z \leq 1.96 \qquad \left(130 - 1.96\,\frac{11.8}{\sqrt{10}},\ 130 + 1.96\,\frac{11.8}{\sqrt{10}}\right)$$

or $\qquad z = \dfrac{X - \mu}{\theta/\sqrt{n}} = -1.96 \qquad$ (122.7, 137.3).

b. The interval may be described in one of the following ways: we are 95% confident that this interval covers the true mean systolic blood pressure μ_s, or there is a 95% chance that this interval covers μ_s before a sample is selected, or approximately 95 out of 100 intervals constructed in this way will cover μ_s.

c. A two-sided 90% confidence interval for μ_d is

$$\left(84 - 1.645\,\frac{9.1}{\sqrt{10}},\ 84 + 1.645\,\frac{9.1}{\sqrt{10}}\right)$$

or

$$(79.3,\ 88.7).$$

d. A two-sided 99% confidence interval for μ_d is

$$\left(84 - 2.58\,\frac{9.1}{\sqrt{10}},\ 84 + 2.58\,\frac{9.1}{\sqrt{10}}\right)$$

or

$$(76.6,\ 91.4).$$

e. The 99% confidence interval is wider than the 90% interval. (The smaller the range of values that is considered, the less confident we are that the interval covers μ_d.)

Exercise 7

a. For the t distribution with 21 degrees of freedom, 1% of the area lies to the left of $t = -2.518$.

b. 10% of the area lies to the right of $t = 1.323$.

c. Since 5% of the area lies to the left of $t = -1.721$ and another 0.5% lies to the right of $t = 2.831$, 94.5% of the area lies between the two values.

d. The value $t = -2.080$ cuts off the lower 2.5% of the distribution.

Exercise 9

a. Since the population standard deviation σ is unknown, we use the t distribution with 13 df rather than the normal distribution. A two-sided 95% confidence interval for μ is

$$\left(29.6 - 2.160\,\frac{3.6}{\sqrt{14}},\ 29.6 + 2.160\,\frac{3.6}{\sqrt{14}}\right)$$

or

$$(27.5,\ 31.7).$$

b. The length of this interval is $31.7 - 27.5 = 4.2$ weeks.

c. Since the interval is centered around the sample mean $\bar{x} = 29.6$ weeks, we are interested in the sample size necessary to produce the interval

$$(29.6 - 1.5, 29.6 + 1.5)$$

or

$$(28.1, 31.1).$$

We know that the 95% confidence interval is of the form

$$\left(29.6 - 1.96\frac{3.6}{\sqrt{n}}, 29.6 + 1.96\frac{3.6}{\sqrt{n}}\right).$$

To find n, therefore, we must solve the equation

$$1.5 = \frac{1.96(3.6)}{\sqrt{n}}$$

or

$$n = \left[\frac{1.96(3.6)}{1.5}\right]^2$$

$$= 22.1.$$

A sample of size 23 is required.

d. Here we are interested in the sample size necessary to produce the interval

$$(29.6 - 1, 29.6 + 1)$$

or

$$(28.6, 30.6).$$

The 95% confidence interval takes the form

$$\left(29.6 - 1.96\frac{3.6}{\sqrt{n}}, 29.6 + 1.96\frac{3.6}{\sqrt{n}}\right).$$

To find n, therefore, we solve the equation

$$1 = \frac{1.96(3.6)}{\sqrt{n}}$$

or

$$n = \left[\frac{1.96(3.6)}{1}\right]^2$$

$$= 49.8.$$

A sample of size 50 is required.

Exercise 11

a. Because the population standard deviation is unknown, we use the t distribution with 7 df rather than the normal distribution. The sample mean calcium level is $\bar{x}_c = 3.14$ mmol/l and the standard deviation is $s_c = 0.51$ mmol/l. A one-sided lower 95% confidence bound for the true mean calcium level μ_c is $3.14 - 1.895(0.51/\sqrt{8}) = 2.80$ mmol/l.

b. The sample mean albumin level is $\bar{x}_a = 40.4$ g/l and the standard deviation is $s_a = 3.0$ g/l. A one-sided lower 95% confidence bound for the true mean albumin level μ_a is $40.4 - 1.895(3.0/\sqrt{8}) = 38.4$ g/l.

c. The lower 95% confidence bound for the mean calcium level does not lie within the normal range of values; this suggests that calcium levels are elevated for this group. There is no evidence that albumin levels differ from the normal range.

Exercise 13

a. A 95% confidence interval for the true mean systolic blood pressure of male low birth weight infants is $(44.3, 51.5)$.

```
. ci sbp if sex==1

Variable |     Obs        Mean    Std. Err.      [95% Conf. Interval]
---------+-----------------------------------------------------------
    sbp  |      44    47.86364    1.779788       44.27435    51.45292
```

b. A 95% confidence interval for the true mean systolic blood pressure of female low birth weight infants is $(43.5, 49.4)$.

```
. ci sbp if sex==0

Variable |     Obs        Mean    Std. Err.      [95% Conf. Interval]
---------+-----------------------------------------------------------
    sbp  |      56    46.46429    1.489348       43.47956    49.44901
```

c. It is possible that males and females have the same mean systolic blood pressure. There is a great deal of overlap between the two confidence intervals.

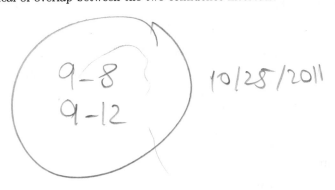

9-8
9-12

10/25/2011

Exercise 9

a. The null hypothesis of the test is

$$H_0 : \mu = 74.4 \text{ mm Hg}.$$

b. The alternative hypothesis is

$$H_A : \mu \neq 74.4 \text{ mm Hg}.$$

c. The test statistic is

$$
\begin{aligned}
z &= \frac{\bar{x}_d - \mu_0}{\sigma_d/\sqrt{n}} \\
&= \frac{84 - 74.4}{9.1/\sqrt{10}} \\
&= 3.34.
\end{aligned}
$$

The area to the right of $z = 3.34$ is less than 0.001, and the area to the left of $z = -3.34$ is less than 0.001 as well; therefore, $p < 0.002$.

d. Since $p < 0.05$, we reject H_0 and conclude that the mean diastolic blood pressure for the population of female diabetics between the ages of 30 and 34 is not equal to 74.4 mm Hg. In fact, it is higher.

e. Since $p < 0.01$, the conclusion would have been the same.

$10-10$

Exercise 11 $b \, \& \, c$

a. Since the population standard deviation is unknown, we use the t distribution with $58 - 1 = 57$ df rather than the normal. A t distribution with 57 df can be approximated by a t distribution with 60 df; in this case, 95% of the observations lie between -2.000 and 2.000. (More accurately, if df $= 57$ then 95% of the observations lie between -2.002 and 2.002.) A two-sided 95% confidence interval for μ is

$$\left(25.0 - 2.000\frac{2.7}{\sqrt{58}}, \ 25.0 + 2.000\frac{2.7}{\sqrt{58}} \right)$$

or

$$(24.3, \ 25.7).$$

b. The null hypothesis for this test is

$$H_0 : \mu = 24.0 \text{ kg/m}^2$$

and the alternative hypothesis is

$$H_A : \mu \neq 24.0 \text{ kg/m}^2.$$

The test statistic is

$$
\begin{aligned}
t &= \frac{\bar{x} - \mu_0}{s/\sqrt{n}} \\
&= \frac{25.0 - 24.0}{2.7/\sqrt{58}} \\
&= 2.82.
\end{aligned}
$$

For a t distribution with 57 degrees of freedom, $2(0.0005) < p < 2(0.005)$ or $0.001 < p < 0.01$. Therefore, we reject H_0.

c. We conclude that the mean baseline body mass index for the population of men who later develop diabetes mellitus is not equal to 24.0 kg/m^2, the mean for the population of men who do not. In fact, it is higher.

d. Since the value 24.0 does not lie inside the 95% confidence interval for μ, we should have expected that the null hypothesis would be rejected.

Exercise 13

It would be impossible for the FDA to completely eliminate the occurrence of type II errors. The probability of committing a type II error is the probability of failing to reject the null hypothesis when it is false; the only way to make this probability equal to 0 is to **always** reject every null hypothesis.

Exercise 15

Since $\alpha = 0.05$, H_0 would be rejected for $z \leq -1.645$. Writing

$$
\begin{aligned}
z &= -1.645 \\
&= \frac{\bar{x} - 3500}{430/\sqrt{n}}
\end{aligned}
$$

and solving for $\bar{x}$,

$$
\bar{x} = 3500 - 1.645 \left(\frac{430}{\sqrt{n}} \right).
$$

The null hypothesis would be rejected for this value. The value of z that corresponds to $\beta = 0.10$ for a two-sided test is 1.645; for the distribution centered at $\mu_1 = 3200$ grams,

$$
1.645 = \frac{\bar{x} - 3200}{430/\sqrt{n}}
$$

and

$$
\bar{x} = 3200 + 1.645 \left(\frac{430}{\sqrt{n}} \right).
$$

Equating the two expressions for $\bar{x}$,

$$
\begin{aligned}
n &= \left[\frac{(1.645 + 1.645)(430)}{(3500 - 3200)} \right]^2 \\
&= 22.2.
\end{aligned}
$$

A sample of size 23 would be required.

CHAPTER 11

Exercise 5

a. The samples are paired.

b. The null hypothesis is

$$H_0: \mu_{\text{corn}} - \mu_{\text{oats}} = 0$$

and the alternative hypothesis is

$$H_A: \mu_{\text{corn}} - \mu_{\text{oats}} \neq 0.$$

c. Since the data are paired, we begin by calculating the difference in LDL cholesterol levels for each person in the study.

Subject	Difference
1	0.77
2	0.85
3	−0.45
4	−0.26
5	0.30
6	0.86
7	0.60
8	0.62
9	0.31
10	0.72
11	0.09
12	0.16
13	0.41
14	0.10

Note that

$$\bar{d} = 0.363 \text{ mmol/l}$$

and

$$s_d = 0.406 \text{ mmol/l}.$$

Therefore, the test statistic is

$$
\begin{aligned}
t &= \frac{\bar{d} - \delta}{s_d/\sqrt{n}} \\
&= \frac{0.363 - 0}{0.406/\sqrt{14}} \\
&= 3.35.
\end{aligned}
$$

For a t distribution with $14 - 1 = 13$ degrees of freedom, $0.001 < p < 0.01$. We reject H_0 at the 0.05 level of significance.

d. We conclude that the true difference in population mean cholesterol levels (or the true mean difference) is not equal to 0. Mean LDL cholesterol is lower when individuals are adhering to the oat bran diet.

Exercise 7

a. Since the samples of data are paired, we first calculate the difference in saliva cotinine levels for each individual.

Subject	Difference
1	49
2	31
3	18
4	34
5	33
6	7
7	104

Note that

$$\bar{d} \; = \; 39.4 \text{ nmol/l}$$

and

$$s_d \; = \; 31.4 \text{ nmol/l.}$$

For a t distribution with $7 - 1 = 6$ degrees of freedom, 95% of the values lie above -1.943. Therefore, a one-sided 95% confidence interval for the true difference in population means $\delta = \mu_{12} - \mu_{24}$ is

$$\delta \; \geq \; \bar{d} - 1.943 \left(\frac{s_d}{\sqrt{n}} \right)$$

$$= \; 39.4 - 1.943 \left(\frac{31.4}{\sqrt{7}} \right)$$

$$= \; 16.3.$$

b. The null hypothesis is

$$H_0 : \mu_{12} - \mu_{24} \leq 0$$

and the alternative hypothesis is

$$H_A : \mu_{12} - \mu_{24} > 0.$$

Given that $\delta = \mu_{12} - \mu_{24} = 0$, the test statistic is

$$t \; = \; \frac{\bar{d} - \delta}{s_d / \sqrt{n}}$$

$$= \; \frac{39.4 - 0}{31.4 / \sqrt{7}}$$

$$= \; 3.32.$$

Since $0.005 < p < 0.01$, we reject H_0 at the 0.05 level of significance and conclude that the true difference in population mean cotinine levels is not equal to 0. Mean cotinine level decreases significantly between 12 and 24 hours after smoking.

Exercise 9

a. The null hypothesis of the test is

$$H_0: \mu_1 = \mu_2$$

and the alternative hypothesis is

$$H_A: \mu_1 \neq \mu_2.$$

Since $s_1 = s_2 = 8$ mm Hg, the pooled estimate of the variance is

$$s_p^2 = 8^2$$
$$= 64.$$

Furthermore, the test statistic is

$$t = \frac{(\bar{x}_1 - \bar{x}_2) - (\mu_1 - \mu_2)}{\sqrt{s_p^2[(1/n_1) + (1/n_2)]}}$$

$$= \frac{(111 - 109) - 0}{\sqrt{64\,[(1/23) + (1/24)]}}$$

$$= 0.86.$$

For a t distribution with $23 + 24 - 2 = 45$ degrees of freedom, $p > 0.10$. Therefore, we are unable to reject H_0 at the 0.01 level of significance. We do not have any evidence that mean arterial blood pressure differs for the two populations of women.

b. To begin, we can approximate the t distribution with 45 df by the t distribution with 40 df. In this case, 99% of the observations are enclosed by the values -2.704 and 2.704. (In fact, if df $= 45$, then 99% of the observations lie between -2.690 and 2.690.) Therefore, a 99% confidence interval for the true difference in population means $\mu_1 - \mu_2$ is

$$(\bar{x}_1 - \bar{x}_2) \pm 2.704\sqrt{s_p^2\left[\frac{1}{n_1} + \frac{1}{n_2}\right]}$$

or

$$(111 - 109) \pm 2.704\sqrt{64\left[\frac{1}{23} + \frac{1}{24}\right]}$$

or

$$(-4.3\,,\, 8.3).$$

This interval does contain the value 0. Given that we were unable to reject the null hypothesis at the 0.01 level, we should have expected that it would.

Exercise 11

a. The null hypothesis of the test is

$$H_0: \mu_1 \geq \mu_2$$

and the alternative hypothesis is

$$H_A: \mu_1 < \mu_2.$$

b. Since we are unwilling to assume that the population variances are identical, we use the modified two-sample t-test. The test statistic is

$$t = \frac{(\bar{x}_1 - \bar{x}_2) - (\mu_1 - \mu_2)}{\sqrt{(s_1^2/n_1) + (s_2^2/n_2)}}$$

$$= \frac{(1.3 - 4.1) - 0}{\sqrt{(1.3^2/121) + (2.0^2/75)}}$$

$$= -10.79.$$

We now calculate the approximate degrees of freedom. Since $s_1^2 = (1.3)^2 = 1.69$ and $s_2^2 = (2.0)^2 = 4.00$,

$$v = \frac{[(s_1^2/n_1) + (s_2^2/n_2)]^2}{[(s_1^2/n_1)^2/(n_1 - 1) + (s_2^2/n_2)^2/(n_2 - 1)]}$$

$$= \frac{[(1.69/121) + (4.00/75)]^2}{[(1.69/121)^2/(121 - 1) + (4.00/75)^2/(75 - 1)]}$$

$$= 113.1.$$

Rounding down to the nearest integer, $v = 113$. For a t distribution with 113 df, $p < 0.0005$. Therefore, we reject the null hypothesis at the 0.05 level of significance and conclude that the mean carboxyhemoglobin level of the nonsmokers is lower than the mean level of the smokers.

Exercise 13
a. Numerical summary measures for the numbers of community hospital beds in 1980 and 1986 — including the mean, the median, and the minimum and maximum values — appear below.

```
. summarize bed80, detail
```

```
                    beds per 1000 pop in 1980
-------------------------------------------------------------

          Percentiles      Smallest
   1%         2.7             2.7
   5%         3.1             3.1
  10%         3.5             3.1        Obs               51
  25%         3.7             3.1        Sum of Wgt.       51

  50%         4.5                        Mean         4.556863
                            Largest      Std. Dev.    1.012769
  75%         5.1             5.9
  90%         5.7             6          Variance     1.025702
  95%         6               7.3        Skewness      .6143899
  99%         7.4             7.4        Kurtosis      3.45173
```

```
. summarize bed86, detail
```

beds per 1000 pop in 1986

```
-------------------------------------------------------------------
        Percentiles      Smallest
  1%        2.4             2.4
  5%        2.7             2.6
 10%        3.1             2.7        Obs                  51
 25%        3.4             2.9        Sum of Wgt.          51

 50%        4.2                        Mean           4.233333
                         Largest       Std. Dev.      1.107369
 75%          5             5.9
 90%        5.3             6.5        Variance       1.226267
 95%        6.5             7.2        Skewness       .9470785
 99%        7.7             7.7        Kurtosis       4.154782
```

b. The null hypothesis of the test is

$$H_0 : \mu_{80} = \mu_{86}$$

and the alternative hypothesis is

$$H_A : \mu_{80} \neq \mu_{86}.$$

The p-value for the two-sample t-test assuming equal variances is $p = 0.1268 > 0.05$. Therefore, we cannot reject H_0. Using this test, the samples do not provide evidence that the population means are different.

```
. ttest bed80 = bed86, unpaired
```

Two-sample t test with equal variances

Variable	Obs	Mean	Std. Err.	Std. Dev.	[95% Conf. Interval]	
bed80	51	4.556863	.1418161	1.012769	4.272017	4.841709
bed86	51	4.233333	.1550627	1.107369	3.921881	4.544786
combined	102	4.395098	.1057774	1.068299	4.185264	4.604932
diff		.3235294	.2101339		-.0933702	.740429

Degrees of freedom: 100

Ho: mean(bed80) - mean(bed86) = diff = 0

Ha: diff < 0	Ha: diff ~= 0	Ha: diff > 0
t = 1.5396	t = 1.5396	t = 1.5396
P < t = 0.9366	P > |t| = 0.1268	P > t = 0.0634

c. For the paired t-test, $p < 0.0001$. We reject the null hypothesis and conclude that the mean number of community hospital beds per 1000 population in 1986 is not equal to the mean number of beds in 1980; the mean number of beds has decreased over the six-year period.

```
. ttest bed80 = bed86
```

Paired t test

```
------------------------------------------------------------------------------
Variable |   Obs        Mean     Std. Err.    Std. Dev.    [95% Conf. Interval]
---------+--------------------------------------------------------------------
  bed80  |    51     4.556863    .1418161     1.012769     4.272017    4.841709
  bed86  |    51     4.233333    .1550627     1.107369     3.921881    4.544786
---------+--------------------------------------------------------------------
   diff  |    51     .3235294    .0470784     .3362072     .2289696    .4180892
------------------------------------------------------------------------------
```

$$\text{Ho: mean(bed80 - bed86) = mean(diff) = 0}$$

Ha: mean(diff) < 0	Ha: mean(diff) ~= 0	Ha: mean(diff) > 0
t = 6.8721	t = 6.8721	t = 6.8721
P < t = 1.0000	P > |t| = 0.0000	P > t = 0.0000

d. We do not reach the same conclusion using the two different tests. The two-sample t-test assumes that the populations of interest are independent and suggests that the mean number of hospital beds in 1986 is not significantly different from the mean number of beds in 1980. On the other hand, the paired t-test indicates that the mean number of beds has decreased over the six-year period. Pairing helps to control for extraneous sources of variation and should be taken into account when applicable. If pairing is ignored, we might fail to reject a false null hypothesis.

e. The 95% confidence interval for the true difference in means $\mu_{80} - \mu_{86}$ is $(0.23, 0.42)$.

```
. generate diff = bed80 - bed86
. ci diff
```

```
------------------------------------------------------------------------------
Variable |   Obs        Mean     Std. Err.        [95% Conf. Interval]
---------+--------------------------------------------------------------------
   diff  |    51     .3235294    .0470784         .2289696    .4180892
```

Exercise 15

a. The p-value for the two-sample t-test that does not assume equal variances is $p = 0.0264$. At the 0.05 level of significance, we reject the null hypothesis that mean PDI score is identical for the two treatment groups; mean PDI is higher for the low-flow group.

```
. ttest pdi, by(trtment) unequal
```

Two-sample t test with unequal variances

```
------------------------------------------------------------------------------
   Group |   Obs        Mean     Std. Err.    Std. Dev.    [95% Conf. Interval]
---------+--------------------------------------------------------------------
      CA |    73     91.91781    1.929775     16.488       88.07087    95.76474
      LF |    70     97.77143    1.755225     14.68527     94.26985    101.273
---------+--------------------------------------------------------------------
```

```
combined |     143     94.78322    1.325531    15.85104    92.16289    97.40354
---------+-------------------------------------------------------------------
    diff |             -5.85362    2.60861                -11.0109   -.6963372
---------------------------------------------------------------------------
Satterthwaite's degrees of freedom:  140.247

                    Ho: mean(CA) - mean(LF) = diff = 0

    Ha: diff < 0              Ha: diff ~= 0              Ha: diff > 0
      t =  -2.2440              t =  -2.2440              t =  -2.2440
  P < t =   0.0132          P > |t| =   0.0264        P > t =   0.9868
```

b. The p-value for this test is $p = 0.2129$. Here we are unable to reject the null hypothesis that mean MDI score is identical for the two treatment groups.

```
. ttest mdi, by(trtment) unequal

Two-sample t test with unequal variances
---------------------------------------------------------------------------
   Group |    Obs        Mean    Std. Err.   Std. Dev.   [95% Conf. Interval]
---------+-------------------------------------------------------------------
      CA |     74    103.1622    1.914019    16.46501    99.34753    106.9768
      LF |     70       106.4    1.741754    14.57256    102.9253    109.8747
---------+-------------------------------------------------------------------
combined |    144    104.7361    1.300364    15.60437    102.1657    107.3065
---------+-------------------------------------------------------------------
    diff |             -3.237838    2.58789               -8.353798    1.878122
---------------------------------------------------------------------------
Satterthwaite's degrees of freedom:  141.386

                    Ho: mean(CA) - mean(LF) = diff = 0

    Ha: diff < 0              Ha: diff ~= 0              Ha: diff > 0
      t =  -1.2511              t =  -1.2511              t =  -1.2511
  P < t =   0.1065          P > |t| =   0.2129        P > t =   0.8935
```

c. The tests suggest that while treatment group may not affect subsequent mental development, it could in fact have an effect on psychomotor development (with the "circulatory arrest" group having a lower level of development than the "low-flow" group).

CHAPTER 12

Exercise 5

a. For $F_{8,16}$, 10% of the area under the curve lies to the right of $F = 2.09$.

b. The value $F = 3.89$ cuts off the upper 1% of the distribution.

c. Since 0.5% of the area under the curve lies to the right of $F = 4.52$, 99.5% lies to the left of this value.

Exercise 7

a. The estimate of the within-groups variance is

$$s_W^2 = \frac{(n_1 - 1)s_1^2 + (n_2 - 1)s_2^2 + (n_3 - 1)s_3^2 + (n_4 - 1)s_4^2}{n_1 + n_2 + n_3 + n_4 - 4}$$

$$= \frac{(268)(13.4)^2 + (52)(10.1)^2 + (27)(11.6)^2 + (8)(12.2)^2}{269 + 53 + 28 + 9 - 4}$$

$$= 164.1.$$

b. Since the grand mean of the data is

$$\bar{x} = \frac{n_1\bar{x}_1 + n_2\bar{x}_2 + n_3\bar{x}_3 + n_4\bar{x}_4}{n_1 + n_2 + n_3 + n_4}$$

$$= \frac{269(115) + 53(114) + 28(118) + 9(126)}{269 + 53 + 28 + 9}$$

$$= 115.36,$$

the estimate of the between-groups variance is

$$s_B^2 = \frac{n_1(\bar{x}_1 - \bar{x})^2 + n_2(\bar{x}_2 - \bar{x})^2 + n_3(\bar{x}_3 - \bar{x})^2 + n_4(\bar{x}_4 - \bar{x})^2}{4 - 1}$$

$$= \frac{269(-0.36)^2 + 53(-1.36)^2 + 28(2.64)^2 + 9(10.64)^2}{4 - 1}$$

$$= 449.0.$$

c. To test the null hypothesis that the mean systolic blood pressures of the four groups are identical versus the alternative hypothesis that they are not, we calculate the test statistic

$$F = \frac{s_B^2}{s_W^2}$$

$$= \frac{449.0}{164.1}$$

$$= 2.74.$$

For an F distribution with $4 - 1 = 3$ and $359 - 4 = 355$ degrees of freedom, $0.025 < p < 0.050$. Therefore, we reject H_0 at the 0.05 level of significance. We conclude that there is a difference among the mean systolic blood pressures of the four groups.

d. To conduct $\binom{4}{2} = 6$ tests and keep the overall probability of committing a type I error at 0.05, we set the significance level for each individual comparison at

$$\frac{4 \times 3 \times 2 \times 1}{(2!)} = \frac{12}{4} = 2 \qquad \alpha^* = \frac{0.05}{6} \to 12 \qquad \longrightarrow \quad \frac{0.05}{12} = 0.0041$$

$$= 0.0083.$$

To compare the mean systolic blood pressures of nonsmokers and current smokers, we calculate the test statistic

$$t_{12} = \frac{\bar{x}_1 - \bar{x}_2}{\sqrt{s_w^2 \left[(1/n_1) + (1/n_2)\right]}}$$

$$= \frac{115 - 114}{\sqrt{164.1 \left[(1/269) + (1/53)\right]}}$$

$$= 0.52.$$

A t distribution with $359 - 4 = 355$ df can be approximated by the standard normal distribution. In this case, $p > 0.0083$; therefore, we are unable to reject the null hypothesis that the two means are identical. Similarly, we can calculate the appropriate test statistics for each of the other five comparisons.

$$t_{13} = \frac{115 - 118}{\sqrt{164.1 \left[(1/269) + (1/28)\right]}}$$

$$= -1.18$$

$$t_{14} = \frac{115 - 126}{\sqrt{164.1 \left[(1/269) + (1/9)\right]}}$$

$$= -2.53$$

$$t_{23} = \frac{114 - 118}{\sqrt{164.1 \left[(1/53) + (1/28)\right]}}$$

$$= -1.34$$

$$t_{24} = \frac{114 - 126}{\sqrt{164.1 \left[(1/53) + (1/9)\right]}}$$

$$= -2.60$$

$$t_{34} = \frac{118 - 126}{\sqrt{164.1 \left[(1/28) + (1/9)\right]}}$$

$$= -1.63$$

Two of the comparisons result in a p-value that is less than 0.0083 — the comparison of group 1 with group 4, and the comparison of group 2 with group 4. Therefore, we conclude that the mean systolic blood pressures of both nonsmokers and current smokers are lower than the mean systolic blood pressure of tobacco chewers.

Exercise 9

a. The average minutes of individual therapy per session is longest for the private not-for-profit centers and shortest for the for-profit centers; the average minutes of group therapy is longest for the for-profit centers and shortest for the public centers.

b. Each 95% confidence interval takes the form $\bar{x} \pm t_{n-1}\left(s/\sqrt{n}\right)$. The intervals are:

Center	Individual Therapy	Group Therapy
FP	$(44.30, 54.62)$	$(89.81, 121.85)$
NFP	$(53.49, 56.03)$	$(95.10, 102.26)$
Public	$(51.57, 54.93)$	$(90.00, 98.34)$

For individual therapy, the three confidence intervals all overlap. Therefore, there is nothing to suggest that the populations means are not identical. (This is not a formal test, however.) The same is true for group therapy.

c. To test the null hypothesis that the mean minutes of individual therapy per session are identical for each type of center, we first calculate estimates of the within-groups and between-groups variances. Note that

$$
\begin{aligned}
s_W^2 &= \frac{(n_1-1)s_1^2 + (n_2-1)s_2^2 + (n_3-1)s_3^2}{n_1 + n_2 + n_3 - 3} \\
&= \frac{(36)(15.47)^2 + (311)(11.41)^2 + (168)(11.08)^2}{37 + 312 + 169 - 3} \\
&= 135.4.
\end{aligned}
$$

Since

$$
\begin{aligned}
\bar{x} &= \frac{n_1\bar{x}_1 + n_2\bar{x}_2 + n_3\bar{x}_3}{n_1 + n_2 + n_3} \\
&= \frac{37(49.46) + 312(54.76) + 169(53.25)}{37 + 312 + 169} \\
&= 53.88,
\end{aligned}
$$

we have that

$$
\begin{aligned}
s_B^2 &= \frac{n_1(\bar{x}_1 - \bar{x})^2 + n_2(\bar{x}_2 - \bar{x})^2 + n_3(\bar{x}_3 - \bar{x})^2}{3-1} \\
&= \frac{37(-4.42)^2 + 312(0.88)^2 + 169(-0.63)^2}{3-1} \\
&= 515.8.
\end{aligned}
$$

Therefore, the test statistic is

$$F = \frac{s_B^2}{s_W^2}$$

$$= \frac{515.8}{135.4}$$

$$= 3.81.$$

For an F distribution with $3 - 1 = 2$ and $518 - 3 = 515$ df, $0.01 < p < 0.025$. We reject the null hypothesis at the 0.05 level of significance and conclude that mean minutes of individual therapy per session are not the same for all three types of centers.

To apply the Bonferroni method of multiple comparisons, we conduct three pairwise tests at the $0.05/3 = 0.0167$ level of significance. The test statistics are:

$$t_{12} = \frac{49.46 - 54.76}{\sqrt{135.4\,[(1/37) + (1/312)]}}$$

$$= -2.62$$

$$t_{13} = \frac{49.46 - 53.25}{\sqrt{135.4\,[(1/37) + (1/169)]}}$$

$$= -1.79$$

$$t_{23} = \frac{54.76 - 53.25}{\sqrt{135.4\,[(1/312) + (1/169)]}}$$

$$= 1.36$$

All test statistics have a t distribution with 515 df. The comparison of private for-profit and not-for-profit centers results in $p = 0.008$; the p-values for the other two comparisons are both greater than 0.0167. Therefore, we conclude that the mean minutes of individual therapy per session is shorter for for-profit centers than for not-for-profit centers.

d. To test the null hypothesis that the mean minutes of group therapy per session are identical for each type of center, we again calculate estimates of the within-groups and between-groups variances. Note that

$$s_W^2 = \frac{(29)(42.91)^2 + (295)(31.27)^2 + (164)(27.12)^2}{30 + 296 + 165 - 3}$$

$$= 947.7.$$

Since

$$\bar{x} = \frac{30(105.83) + 296(98.68) + 165(94.17)}{30 + 296 + 165}$$

$$= 97.60,$$

we have that

$$s_B^2 = \frac{30(8.23)^2 + 296(1.08)^2 + 165(-3.43)^2}{3 - 1}$$

$$= 2159.2.$$

43

The test statistic is

$$F = \frac{s_B^2}{s_W^2}$$

$$= \frac{2159.2}{947.7}$$

$$= 2.28.$$

For an F distribution with $3 - 1 = 2$ and $491 - 3 = 488$ df, $p > 0.10$. Therefore, we are unable to reject the null hypothesis that mean minutes of group therapy per session are identical.
e. While private for-profit centers have shorter individual therapy sessions than not-for-profit centers, on average, there are no significant differences in the length of group therapy sessions.

Exercise 11
a. Using the two-sample t test that assumes equal variances, the test statistic is t$=-0.6072$ and $p = 0.5451$. We are unable to reject the null hypothesis that mean systolic blood pressure is identical for girls and boys.

```
. ttest sbp, by(sex)

Two-sample t test with equal variances

-----------------------------------------------------------------------------
  Group |     Obs        Mean    Std. Err.   Std. Dev.   [95% Conf. Interval]
--------+--------------------------------------------------------------------
 Female |      56    46.46429    1.489348    11.14526    43.47956    49.44901
   Male |      44    47.86364    1.779788    11.80577    44.27435    51.45292
--------+--------------------------------------------------------------------
combined |    100       47.08    1.140324    11.40324    44.81735    49.34265
--------+--------------------------------------------------------------------
   diff |           -1.399351    2.304609               -5.972771     3.17407
-----------------------------------------------------------------------------
Degrees of freedom: 98

              Ho: mean(Female) - mean(Male) = diff = 0

   Ha: diff < 0                 Ha: diff ~= 0                 Ha: diff > 0
      t =  -0.6072                 t =  -0.6072                 t =  -0.6072
  P < t =   0.2726           P > |t| =   0.5451             P > t =   0.7274
```

b. Using the one-way analysis of variance, the test statistic is F$=0.37$ and $p = 0.5451$. Again we are unable to reject the null hypothesis that mean systolic blood pressure is identical for girls and boys.

```
. oneway sbp sex
```

```
                          Analysis of Variance
        Source              SS          df       MS              F      Prob > F
-----------------------------------------------------------------------------
Between groups           48.2496104      1    48.2496104        0.37      0.5451
Within groups            12825.1104     98    130.868473
-----------------------------------------------------------------------------
        Total            12873.36       99    130.033939

Bartlett's test for equal variances:  chi2(1) =    0.1590  Prob>chi2 = 0.690
```

c. The F-test and the t-test do appear to be mathematically equivalent. Note that the p-values are exactly the same.

CHAPTER 13

Exercise 7

a. We want to evaluate the null hypothesis that the median difference in respiratory rates is equal to 0, using the sign test.

Pair	Difference	Sign	Rank	Signed Rank
1	16	+	10	10
2	−7	−	5	−5
3	−2	−	1.5	−1.5
4	38	+	13	13
5	12	+	7	7
6	2	+	1.5	1.5
7	23	+	11	11
8	−14	−	9	−9
9	6	+	4	4
10	−13	−	8	−8
11	−3	−	3	−3
12	36	+	12	12
13	8	+	6	6
14	40	+	14	14

Because $n < 20$, we must use the small sample version of the test. If the null hypothesis is true, the number of plus signs D has a binomial distribution with parameters $n = 14$ and $p = 0.5$. We would expect to see $14(0.5) = 7$ plus signs in the sample; we observe $D = 9$. The probability of observing 9 or more plus signs is

$$
\begin{aligned}
P(D \geq 9) &= P(D = 9) + P(D = 10) + P(D = 11) + P(D = 12) + P(D = 13) + P(D = 14) \\
&= 0.1222 + 0.0611 + 0.0222 + 0.0056 + 0.0009 + 0.0001 \\
&= 0.2121.
\end{aligned}
$$

This is the p-value of the one-sided test; the p-value of the corresponding two-sided test is approximately $p = 2(0.2121) = 0.4242$. We are unable to reject the null hypothesis at the 0.05 level of significance, and thus have no evidence that the median difference in respiratory rates is different from 0.

b. We now use the signed-rank test. The sum of the positive ranks is 78.5 and the sum of the negative ranks is −26.5; therefore, ignoring the signs, the smaller sum of ranks is $T = 26.5$. The mean sum of ranks is

$$
\begin{aligned}
\mu_T &= \frac{n(n+1)}{4} \\
&= \frac{14(15)}{4} \\
&= 52.5
\end{aligned}
$$

and the standard deviation is

$$\sigma_T = \sqrt{\frac{n(n+1)(2n+1)}{24}}$$

$$= \sqrt{\frac{14(15)(29)}{24}}$$

$$= 15.9.$$

Solving for the test statistic z_T,

$$z_T = \frac{T - \mu_T}{\sigma_T}$$

$$= \frac{26.5 - 52.5}{15.9}$$

$$= -1.64.$$

Therefore, $p = 2(0.051) = 0.102$. Since $p > 0.05$, we are again unable to reject the null hypothesis.

c. We do reach the same conclusion in each case.

Exercise 9

a. If $T = 2$, the p-value of the two-sided test is approximately $p = 2(0.0469) = 0.0938$.

b. In this case, $T = 6$. The p-value of the two-sided test is approximately $p = 2(0.2188) = 0.4376$.

Exercise 11

a. To test the null hypothesis that the median age at death is the same for males as it is for females, we first combine the subjects into a single group and rank the observations from smallest to largest.

Females		Males			
Age	Rank	Age	Rank	Age	Rank
53	3.0	46	1.0	115	19.0
56	4.0	52	2.0	133	21.0
60	7.5	58	5.0	134	22.5
60	7.5	59	6.0	167	25.0
78	10.5	77	9.0	175	26.0
87	15.0	78	10.5		
102	16.0	80	12.0		
117	20.0	81	13.0		
134	22.5	84	14.0		
160	24.0	103	17.0		
277	27.0	114	18.0		

The smaller sum of ranks, $W = 157$, corresponds to the group of females. The mean sum of the ranks is

$$\mu_W = \frac{n_S(n_S + n_L + 1)}{2}$$

$$= \frac{11(11 + 16 + 1)}{2}$$

$$= 154$$

and the standard deviation is

$$\sigma_W = \sqrt{\frac{n_S n_L (n_S + n_L + 1)}{12}}$$

$$= \sqrt{\frac{11(16)(11 + 16 + 1)}{12}}$$

$$= 20.3.$$

The test statistic is

$$z_W = \frac{W - \mu_W}{\sigma_W}$$

$$= \frac{157 - 154}{20.3}$$

$$= 0.15.$$

Therefore, $p > 0.10$ and we do not reject the null hypothesis. There is no evidence that the median age at death for females is different from the median age at death for males.

b. Histograms of the ages at death for females and males are shown below. Because the values are skewed to the right, it would not be appropriate to use the two-sample t-test to analyze these data.

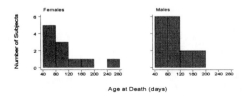

Age at Death (days)

Exercise 13

a. The box plots showing the numbers of community hospital beds in 1980 and 1986 are displayed on the following page. Note that the values appear to be fairly symmetric.

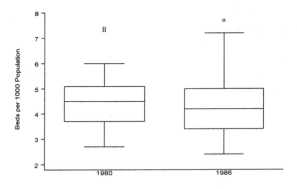

b. When conducting the Wilcoxon signed-rank test, the null hypothesis is that the median difference in the number of community hospital beds per 1000 population is equal to 0. Since $p < 0.0001$, we reject H_0 and conclude that the median difference is not equal to 0.

```
. signrank bed80=bed86

Wilcoxon signed-rank test

       sign |       obs    sum ranks     expected
   ---------+---------------------------------------
   positive |        48      1203.5          663
   negative |         3       122.5          663
       zero |         0           0            0
   ---------+---------------------------------------
        all |        51        1326         1326

unadjusted variance       11381.50
adjustment for ties         -19.12
adjustment for zeros          0.00
                         ----------
adjusted variance         11362.38

Ho: bed80 = bed86
            z =     5.071
    Prob > |z| =    0.0000
```

c. For the Wilcoxon rank sum test, the null hypothesis is that the median number of community hospital beds per 1000 population in 1980 is equal to the median number of beds in 1986. In this case, $p = 0.065$. If we are conducting the test at the 0.05 level of significance, we would be unable to reject the null hypothesis that the medians are the same.

```
. ranksum bed, by(year)

Two-sample Wilcoxon rank-sum (Mann-Whitney) test

    year |       obs     rank sum      expected
---------+------------------------------------
    1980 |        51         2902        2626.5
    1986 |        51         2351        2626.5
---------+------------------------------------
combined |       102         5253          5253

unadjusted variance      22325.25
adjustment for ties        -42.29
                       ----------
adjusted variance        22282.96

Ho: bed(year==1980) = bed(year==1986)
           z =     1.846
    Prob > |z| =    0.0650
```

d. Using the rank sum test, which assumes that the two samples of data are independent, we do not have quite sufficient evidence to conclude that the median number of community hospital beds in 1980 is different from the median number of beds in 1986. When we apply the signed-rank test, which assumes that the samples are paired, we conclude that there were more beds per 1000 population in 1980 than there were six years later. Thus, the tests do not lead us to draw the same conclusions. Pairing helps to minimize extraneous sources of variability and should be taken into account when applicable. Consequently, the signed-rank test is more appropriate for these data.

e. For both the parametric and nonparametric procedures, the same conclusions are drawn. The paired analyses lead us to reject the null hypothesis of no difference, while the analyses which assume independent populations do not. Since the data appear to be approximately normally distributed, it is not surprising that we reach the same conclusions in each case.

Exercise 15

a. The box plots are shown below. The distributions of Apgar scores are very similar for males and females.

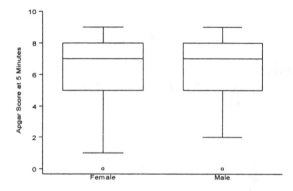

b. We use the rank sum test to evaluate the null hypothesis that median five-minute Apgar scores are equal for males and females. Since $p = 0.75$, we are unable to reject the null hypothesis. There is no evidence that Apgar scores differ by gender.

```
. ranksum apgar5, by(sex)

Two-sample Wilcoxon rank-sum (Mann-Whitney) test

        sex |      obs    rank sum    expected
---------+-----------------------------------
     Female |       56     2783.5        2828
       Male |       44     2266.5        2222
---------+-----------------------------------
   combined |      100       5050        5050

unadjusted variance      20738.67
adjustment for ties       -639.89
                         ----------
adjusted variance        20098.77

Ho: apgar5(sex==Female) = apgar5(sex==Male)
            z =   -0.314
    Prob > |z| =    0.7536
```

14·6

14·7 (a-e)

Exercise 5   14·10   14.15 (optional)

Exercise 5

a. The exact binomial probability that four or fewer of the infants weigh at most 2500 grams is

$$P(X \le 4) = P(X = 0) + P(X = 1) + P(X = 2) + P(X = 3) + P(X = 4)$$

$$= \binom{40}{0}(0.15)^0(0.85)^{40} + \binom{40}{1}(0.15)^1(0.85)^{39} + \binom{40}{2}(0.15)^2(0.85)^{38}$$

$$+ \binom{40}{3}(0.15)^3(0.85)^{37} + \binom{40}{4}(0.15)^4(0.85)^{36}$$

$$= 0.263.$$

b. Since $np = 40(0.15) = 6$ and $n(1 - p) = 40(0.85) = 34$ are both greater than 5, we can use the normal approximation to the binomial distribution. Applying the continuity correction, we find that

$$z = \frac{X - np + 0.5}{\sqrt{np(1 - p)}}$$

$$= \frac{4 - (40)(0.15) + 0.5}{\sqrt{40(0.15)(0.85)}}$$

$$= -0.66.$$

P = 0·263

P = 0·255

The area under the standard normal curve that lies to the left of $z = -0.66$ is 0.255; this is the estimated probability that at most four of the newborns weigh at most 2500 grams.

c. The normal approximation provides a fairly good estimate of the exact binomial probability.

Exercise 7

a. A point estimate for p is

a - e

$$\hat{p} = \frac{15}{27}$$

$$= 0.556.$$

Since $n\hat{p} = 27(0.556) = 15$ and $n(1 - \hat{p}) = 27(0.444) = 12$, the sample size is large enough to justify the use of the normal approximation. Therefore, an approximate 95% confidence interval for p is

$$\left(0.556 - 1.96 \sqrt{\frac{0.556(1 - 0.556)}{27}}, \ 0.556 + 1.96 \sqrt{\frac{0.556(1 - 0.556)}{27}}\right)$$

or

$$(0.369, 0.743).$$

We are 95% confident that these limits cover the true population proportion p.

b. The null hypothesis of the test is

$$H_0: \mu = 0.328.$$

c. The alternative hypothesis is

$$H_A: \mu \neq 0.328.$$

d. The test statistic is

$$z = \frac{\hat{p} - p}{\sqrt{p(1-p)/n}}$$

$$= \frac{0.556 - 0.328}{\sqrt{0.328(1 - 0.328)/27}}$$

$$= 2.52.$$

Therefore, $p = 2(0.006) = 0.012$. Since $p > 0.01$, we are unable to reject the null hypothesis.

e. We conclude that for children with an oral cleft, there is no evidence that the proportion of mothers who smoked during pregnancy is different from the proportion of mothers who smoked for children with other types of malformations. (Note: If the test were being conducted at the 0.05 level of significance, we would reject H_0 and conclude that the proportion is higher than 32.8%.)

f. In this case, $p_0 = 0.328$ and $p_1 = 0.250$. Since $\alpha = 0.01$ for a two-sided test and $\beta = 0.10$, we have that $z_{\alpha/2} = 2.58$ and $z_\beta = 1.28$, and

$$n = \left[\frac{2.58\sqrt{p_0(1-p_0)} + 1.28\sqrt{p_1(1-p_1)}}{p_1 - p_0} \right]^2$$

$$= \left[\frac{2.58\sqrt{0.328(1-0.328)} + 1.28\sqrt{0.250(1-0.250)}}{0.250 - 0.328} \right]^2$$

$$= 512.3.$$

A sample of size 513 would be required.

Exercise 9

a. The estimated proportion of children whose mothers have had more than 12 years of schooling is

$$\hat{p} = \frac{4}{45}$$

$$= 0.09.$$

Note that $n\hat{p} = 45(0.09) = 4$ and $n(1 - \hat{p}) = 45(0.91) = 41$. Since one of these products is less than 5, we should not use the normal approximation to generate a 90% confidence interval; instead, we should construct an exact binomial interval. If we proceed with the approximate method anyway — knowing that it might not provide adequate results — an "approximate" 90% confidence interval for p is

$$\left(0.09 - 1.645\sqrt{\frac{0.09(1-0.09)}{45}}, \; 0.09 + 1.645\sqrt{\frac{0.09(1-0.09)}{45}} \right)$$

or

$$(0.02, 0.16).$$

b. The null hypothesis of the two-sided test is

$$H_0: \mu = 0.22$$

and the alternative hypothesis is
$$H_A : \mu \neq 0.22.$$

c. Assuming that we can use the normal approximation (which we have already noted is not the case), the test statistic is

$$
\begin{aligned}
z &= \frac{\hat{p} - p}{\sqrt{p(1-p)/n}} \\
&= \frac{0.09 - 0.22}{\sqrt{0.22(1 - 0.22)/45}} \\
&= -2.11.
\end{aligned}
$$

Therefore, the p-value is approximately $p = 2(0.017) = 0.034$. Since $p < 0.05$, we reject the null hypothesis.

d. We conclude that the proportion of children with special educational needs whose mothers have had more than 12 years of schooling is not equal to 0.22; in fact, it is lower.

e. In this case, $p_0 = 0.22$ and $p_1 = 0.10$. Since $\alpha = 0.05$ for a two-sided test and $\beta = 0.05$, we have that $z_{\alpha/2} = 1.96$ and $z_\beta = 1.645$, and

$$
\begin{aligned}
n &= \left[\frac{1.96\sqrt{p_0(1-p_0)} + 1.645\sqrt{p_1(1-p_1)}}{p_1 - p_0} \right]^2 \\
&= \left[\frac{1.96\sqrt{0.22(1-0.22)} + 1.645\sqrt{0.10(1-0.10)}}{0.10 - 0.22} \right]^2 \\
&= 118.3.
\end{aligned}
$$

A sample of size 119 would be required.

Exercise 11

a. For individuals assigned to the prepaid plan, the estimated proportion of patients who visited a community crisis center is

$$
\begin{aligned}
\hat{p}_1 &= \frac{13}{311} \\
&= 0.042.
\end{aligned}
$$

Among those receiving traditional Medicaid,

$$
\begin{aligned}
\hat{p}_2 &= \frac{22}{310} \\
&= 0.071.
\end{aligned}
$$

b. The null hypothesis of the test is
$$H_0 : p_1 = p_2$$

and the alternative hypothesis is
$$H_A : p_1 \neq p_2.$$

The pooled estimate of the common proportion is

$$\hat{p} = \frac{x_1 + x_2}{n_1 + n_2}$$

$$= \frac{13 + 22}{311 + 310}$$

$$= 0.056.$$

Therefore, the test statistic is

$$z = \frac{(\hat{p}_1 - \hat{p}_2) - (p_1 - p_2)}{\sqrt{\hat{p}(1 - \hat{p})[(1/n_1) + (1/n_2)]}}$$

$$= \frac{(0.042 - 0.071) - 0}{\sqrt{0.056(1 - 0.056)[(1/311) + (1/310)]}}$$

$$= -1.57.$$

In this case, $p = 2(0.058) = 0.116$; we are unable to reject the null hypothesis at the 0.10 level of significance.

c. There is insufficient evidence to conclude that the proportions of patients visiting a community crisis center are not identical for those on the prepaid medical plan and those receiving traditional Medicaid.

Exercise 13

a. The estimated proportion of low birth weight infants whose mothers experienced toxemia is 0.21, or 21%.

```
. tabulate tox

   toxemia |
 diagnosis |
for mother |      Freq.      Percent        Cum.
-----------+-----------------------------------
        No |         79        79.00       79.00
       Yes |         21        21.00      100.00
-----------+-----------------------------------
     Total |        100       100.00
```

b. A 95% confidence interval for the true population proportion p is $(0.135, 0.303)$.

```
. ci tox, bin

                                             -- Binomial Exact --
Variable |       Obs        Mean    Std. Err.    [95% Conf. Interval]
---------+----------------------------------------------------------
     tox |       100         .21     .0407308    .1349414    .3029156
```

c. This is an exact binomial interval. (Answers may differ, depending on statistical software used.)

CHAPTER 15

Exercise 7

a. For the chi-square distribution with 17 df, 1.0% of the area under the curve lies to the right of $\chi^2 = 33.41$.

b. About $100\% - 5\% = 95\%$ of the area lies to the left of $\chi^2 = 27.59$.

c. The value $\chi^2 = 24.77$ cuts off the upper 10% of the distribution.

Exercise 9

a. The proportion of subjects who withdrew from the study in the calcitriol group is $27/314 = 0.086$, while the proportion who withdrew in the calcium group is $20/308 = 0.065$.

b. To test the null hypothesis that there is no association between treatment group and withdrawal from the study, we use the chi-square test. To carry out the test, we first calculate the table of expected counts.

	Withdrawal		
Treatment	Yes	No	Total
Calcitriol	23.7	290.3	314
Calcium	23.3	284.7	308
Total	47	575	622

The test statistic is

$$
\begin{aligned}
X^2 &= \sum_{i=1}^{4} \frac{(|O_i - E_i| - 0.5)^2}{E_i} \\
&= \frac{(2.8)^2}{23.7} + \frac{(2.8)^2}{290.3} + \frac{(2.8)^2}{23.3} + \frac{(2.8)^2}{284.7} \\
&= 0.33 + 0.03 + 0.34 + 0.03 \\
&= 0.73.
\end{aligned}
$$

For a chi-square distribution with $(r-1)(c-1) = (2-1)(2-1) = 1$ degree of freedom, $p > 0.10$. Therefore, we are unable to reject H_0 at the 0.05 level of significance. This data does not provide evidence that the proportions of subjects withdrawing from the study differ by treatment group.

Exercise 11

a. To determine whether the results are homogeneous across studies, we perform the chi-square test. Therefore, we first calculate the table of expected counts.

Date of Study	Certificate Status			Total
	Confirmed Accurate	Inaccurate No Change	Incorrect Recoding	
1955–1965	1895.9	398.4	439.6	2734
1970	178.2	37.5	41.3	257
1970–1971	265.6	55.8	61.6	383
1975–1977	798.9	167.9	185.2	1152
1977–1978	398.7	83.8	92.5	575
1980	188.6	39.6	43.7	272
Total	3726	783	864	5373

The corresponding test statistic is

$$X^2 = \sum_{i=1}^{18} \frac{(O_i - E_i)^2}{E_i}$$

$$= \frac{(144.1)^2}{1895.9} + \frac{(-31.4)^2}{398.4} + \frac{(-112.6)^2}{439.6} + \frac{(-29.2)^2}{178.2} + \frac{(22.5)^2}{37.5}$$

$$+ \frac{(6.7)^2}{41.3} + \frac{(22.4)^2}{265.6} + \frac{(-0.8)^2}{55.8} + \frac{(8.4)^2}{61.6} + \frac{(-95.9)^2}{798.9}$$

$$+ \frac{(29.1)^2}{167.9} + \frac{(66.8)^2}{185.2} + \frac{(26.3)^2}{398.7} + \frac{(-21.8)^2}{83.8} + \frac{(-4.5)^2}{92.5}$$

$$+ \frac{(-67.6)^2}{188.6} + \frac{(32.4)^2}{39.6} + \frac{(35.3)^2}{43.7}$$

$$= 209.2.$$

[handwritten: −308 with line pointing to the (−112.6)² term; the (−0.8)² term is circled]

For a chi-square random variable with $(r-1)(c-1) = (6-1)(3-1) = 10$ df, $p < 0.001$. Therefore, we reject the null hypothesis and conclude that the results are not homogeneous across studies.

b. Among deaths which require autopsies, it seems likely that there would be a higher proportion of certificates that contain inaccuracies or require recoding. Therefore, if we use the results of these studies to make inference about the population as a whole, there is a good chance that we will overestimate the proportion of certificates that are not accurate.

Exercise 13

a. To test the null hypothesis that there is no association between retirement status and cardiac arrest, we use McNemar's test. The test statistic is

$$X^2 = \frac{[|r - s| - 1]^2}{r + s}$$

$$= \frac{[|12 - 20| - 1]^2}{12 + 20}$$

$$= 1.53.$$

For a chi-square distribution with 1 df, $p > 0.10$. Therefore, we cannot reject the null hypothesis.

b. The samples do not provide evidence of an association between retirement status and cardiac arrest.

c. The estimated odds of being retired for healthy individuals versus those who have experienced cardiac arrest is

$$\widehat{OR} = \frac{r}{s}$$

$$= \frac{12}{20}$$

$$= 0.6.$$

d. An approximate 95% confidence interval for the natural logarithm of the odds ratio takes the form

$$\ln(\widehat{OR}) \pm 1.96\, \widehat{se}[\ln(\widehat{OR})].$$

Since $\ln(0.6) = -0.511$ and

$$\widehat{se}[\ln(\widehat{OR})] = \sqrt{\frac{r+s}{rs}}$$

$$= \sqrt{\frac{12+20}{12(20)}}$$

$$= 0.365,$$

a 95% confidence interval for $\ln(OR)$ is

$$(-0.511 - 1.96(0.365), -0.511 + 1.96(0.365))$$

or

$$(-1.23, 0.204).$$

Therefore, a 95% confidence interval for the odds ratio itself is

$$(e^{-1.23}, e^{0.204})$$

or

$$(0.29, 1.23).$$

Exercise 15
a. To test the null hypothesis that there is no association between exposure to air pollutants and the occurrence of headaches, we use McNemar's test. The test statistic is

$$X^2 = \frac{[|r-s|-1]^2}{r+s}$$

$$= \frac{[|2-8|-1]^2}{2+8}$$

$$= 2.50.$$

For a chi-square distribution with 1 df, $p > 0.10$. Therefore, we cannot reject the null hypothesis.
b. The samples do not provide evidence of an association between exposure to air pollutants and headaches.

Exercise 17

a. The 2 × 2 contingency table for these data appears below.

PID	Ectopic Pregnancy Yes	No	Total
Yes	28	6	34
No	251	273	524
Total	279	279	558

b. The estimated relative odds of suffering an ectopic pregnancy for women who have had pelvic inflammatory disease versus women who have not is

$$\widehat{OR} = \frac{(28)(273)}{(6)(251)}$$

$$= 5.08.$$

c. The logarithm of the estimated odds ratio is

$$\ln(\widehat{OR}) = \ln(5.08)$$

$$= 1.625,$$

and the estimated standard error of $\ln(\widehat{OR})$ is

$$\widehat{se}[\ln(\widehat{OR})] = \sqrt{\frac{1}{28} + \frac{1}{6} + \frac{1}{251} + \frac{1}{273}}$$

$$= 0.210.$$

Therefore, a 99% confidence interval for the logarithm of the odds ratio takes the form

$$(1.625 - 2.58(0.210), \ 1.625 + 2.58(0.210))$$

or

$$(1.083, 2.167),$$

and a 99% confidence interval for the odds ratio itself is

$$(e^{1.083}, e^{2.167})$$

or

$$(2.95, 8.73).$$

Exercise 19

a. Among women who have used drugs intravenously, 44.8% are HIV-positive. Among those who have not, 8.0% are HIV-positive.

```
. tabulate hiv ivdu, col

           | intravenous drug use
HIV status |      yes         no |      Total
-----------+--------------------+----------
  positive |       61         27 |         88
           |    44.85       7.96 |      18.53
-----------+--------------------+----------
  negative |       75        312 |        387
           |    55.15      92.04 |      81.47
-----------+--------------------+----------
     Total |      136        339 |        475
           |   100.00     100.00 |     100.00
```

b. To test the null hypothesis that there is no association between history of intravenous drug use and HIV seropositivity, we use the chi-square test. Since $p < 0.001$, we reject the null hypothesis at the 0.05 level.

```
. tabulate hiv ivdu, col chi2

           | intravenous drug use
HIV status |      yes         no |      Total
-----------+--------------------+----------
  positive |       61         27 |         88
           |    44.85       7.96 |      18.53
-----------+--------------------+----------
  negative |       75        312 |        387
           |    55.15      92.04 |      81.47
-----------+--------------------+----------
     Total |      136        339 |        475
           |   100.00     100.00 |     100.00

        Pearson chi2(1) =  87.5018   Pr = 0.000
```

c. We conclude that there is an association between intravenous drug use and HIV status; the proportion of women who are seropositive is higher among individuals who have a history of intravenous drug use than among those who do not.

d. The estimated relative odds of being HIV-positive for women who have used intravenous drugs versus those who have not is

$$\widehat{OR} = \frac{(61)(312)}{(75)(27)} = 9.4.$$

Exercise 21

a. Note that when using the generic questionnaire, $151/237 = 64\%$ of individuals were classified as beer drinkers. When using the questionnaire targeting alcohol use, $116/237 = 49\%$ were classified as beer drinkers.

```
. tabulate genques

    generic |
questionnai |
         re |      Freq.      Percent        Cum.
------------+-----------------------------------
    nodrink |         86        36.29       36.29
      drink |        151        63.71      100.00
------------+-----------------------------------
      Total |        237       100.00
```

```
. tabulate alcques

    alcohol |
questionnai |
         re |      Freq.      Percent        Cum.
------------+-----------------------------------
    nodrink |        121        51.05       51.05
      drink |        116        48.95      100.00
------------+-----------------------------------
      Total |        237       100.00
```

To test the null hypothesis that there is no association between drinking status and the type of questionnaire, we use McNemar's test.

```
. mcc genques alcques

                    | Controls               |
Cases               |  Exposed   Unexposed   |      Total
--------------------+------------------------+----------
          Exposed   |      112          39   |        151
        Unexposed   |        4          82   |         86
--------------------+------------------------+----------
            Total   |      116         121   |        237

McNemar's chi2(1) =      28.49        Pr>chi2 = 0.0000
Exact McNemar significance probability       = 0.0000

Proportion with factor
        Cases         .6371308
        Controls      .4894515        [95% conf. interval]
                      ---------        --------------------
        difference    .1476793         .0925942    .2027644
        ratio         1.301724         1.181257    1.434476
        rel. diff.    .2892562         .1997087    .3788037

        odds ratio    9.75             3.517898    37.56536   (exact)
```

Note that $p < 0.0001$; therefore, we reject the null hypothesis of no association.

b. We conclude that an association does exist; if beer-drinking classification differs on the two questionnaires, an individual is more likely to be classified as a beer drinker on the generic questionnaire than on the one that targets alcohol use.

CHAPTER 16

Exercise 5

a. For women who have had at most one sexual partner, the estimated odds ratio is

$$\widehat{OR}_1 = \frac{(12)(118)}{(25)(21)}$$

$$= 2.70.$$

b. For women who have had two or more sexual partners, the estimated odds ratio is

$$\widehat{OR}_2 = \frac{(96)(150)}{(92)(142)}$$

$$= 1.10.$$

c. Within each stratum, the odds of being diagnosed with cervical cancer are higher for women who smoke than for those who do not. This difference appears to be greater for women who have had at most one sexual partner.

d. If the odds ratios are not constant across strata, it does not make sense to compute a single summary value. Furthermore, even if the odds ratios are similar, the number of sexual partners may be a confounder in the relationship between smoking status and cervical cancer. In this case, summing the entries in the tables could lead to Simpson's paradox; the odds ratio for the summed table may be very different from the odds ratios for either of the individual stratum tables.

e. For the test of homogeneity, the null hypothesis is that the relative odds of developing cervical cancer for smokers versus nonsmokers are the same for women who have had zero or one sexual partner as they are for women who have had two or more partners. Equivalently, the null hypothesis is

$$H_0 : OR_1 = OR_2.$$

Since

$$y_1 = \ln(\widehat{OR}_1)$$
$$= \ln(2.70)$$
$$= 0.993,$$

$$y_2 = \ln(\widehat{OR}_2)$$
$$= \ln(1.10)$$
$$= 0.095,$$

$$w_1 = \frac{1}{[(1/12) + (1/25) + (1/21) + (1/118)]}$$
$$= 5.57$$

and

$$w_2 = \frac{1}{[(1/96) + (1/92) + (1/142) + (1/150)]}$$
$$= 28.58,$$

the weighted average Y is

$$Y = \frac{\sum_{i=1}^{2} w_i y_i}{\sum_{i=1}^{2} w_i}$$

$$= \frac{(5.57)(0.993) + (28.58)(0.095)}{5.57 + 28.58}$$

$$= 0.241.$$

The test statistic is

$$X^2 = \sum_{i=1}^{2} w_i (y_i - Y)^2$$

$$= (5.57)(0.993 - 0.241)^2 + (28.58)(0.095 - 0.241)^2$$

$$= 3.76.$$

For a chi-square distribution with 1 df, $0.05 < p < 0.10$. Therefore, if we are conducting the test of homogeneity at the 0.05 level of significance, we do not reject H_0 (just barely). We may proceed with the Mantel-Haenszel method. (Because this was a borderline test result, the Mantel-Haenszel method might not be the best course of action.)
f. The Mantel-Haenszel estimate of the summary odds ratio is

$$\widehat{OR}_{MH} = \frac{\sum_{i=1}^{2}(a_i d_i / T_i)}{\sum_{i=1}^{2}(b_i c_i / T_i)}$$

$$= \frac{(12)(118)/176 + (96)(150)/480}{(25)(21)/176 + (92)(142)/480}$$

$$= 1.26.$$

(Note: In Review Exercise 16 in Chapter 15, the estimate of the crude odds ratio was $\widehat{OR} = 1.52$. The Mantel-Haenszel estimate is a bit lower than the crude estimate, suggesting that, to some extent, the number of sexual partners a woman has had is a confounder in the relationship between smoking status and cervical cancer.)
g. Before constructing a confidence interval, we must first ensure that the strata sizes are large enough to justify the use of the normal approximation. Since

$$\sum_{i=1}^{2} \frac{M_{1i} N_{1i}}{T_i} = 6.9 + 93.2$$

$$= 100.1,$$

$$\sum_{i=1}^{2} \frac{M_{1i} N_{2i}}{T_i} = 26.1 + 144.8$$

$$= 170.9,$$

$$\sum_{i=1}^{2} \frac{M_{2i} N_{1i}}{T_i} = 30.1 + 94.8$$

$$= 124.9,$$

and

$$\sum_{i=1}^{2} \frac{M_{2i} N_{2i}}{T_i} = 112.9 + 147.2$$
$$= 260.1$$

are each greater than 5.0, we may proceed with the construction of a confidence interval. We previously found that $Y = 0.241$. Note that

$$\widehat{se}(Y) = \frac{1}{\sqrt{w_1 + w_2}}$$
$$= \frac{1}{\sqrt{5.57 + 28.58}}$$
$$= 0.171.$$

Therefore, a 99% confidence interval for $\ln(\text{OR})$ is

$$(0.241 - 2.58(0.171),\ 0.241 + 2.58(0.171))$$

or

$$(-0.200,\ 0.682),$$

and a 99% confidence interval for the common odds ratio is

$$(e^{-0.200},\ e^{0.682})$$

or

$$(0.82,\ 1.98).$$

This interval does contain the value 1, which suggests that 1 is a plausible value for the odds ratio. An odds ratio of 1 would indicate that there is no association between smoking status and cervical cancer after controlling for the number of sexual partners a woman has had.
h. To test the null hypothesis

$$H_0: \text{OR} = 1$$

against the alternative

$$H_A: \text{OR} \neq 1$$

using the Mantel-Haenszel test of association, we first find that

$$a_1 = 12,$$

$$m_1 = \frac{(33)(37)}{176}$$
$$= 6.9,$$

$$\sigma_1^2 = \frac{(33)(143)(37)(139)}{(176)^2(175)}$$
$$= 4.48,$$

$$a_2 = 96,$$

$$m_2 = \frac{(238)(188)}{480}$$
$$= 93.2,$$

and

$$\sigma_2^2 = \frac{(238)(242)(188)(292)}{(480)^2(479)}$$
$$= 28.65.$$

Therefore, the test statistic is

$$X^2 = \frac{[\sum_{i=1}^2 a_i - \sum_{i=1}^2 m_i]^2}{\sum_{i=1}^2 \sigma_i^2}$$
$$= \frac{[(12+96) - (6.9+93.2)]^2}{4.48 + 28.65}$$
$$= 1.88.$$

For a chi-square distribution with 1 df, $p > 0.10$. Therefore, we are unable to reject the null hypothesis at the 0.01 level of significance. After adjusting for the number of sexual partners a woman has had, there does not appear to be an association between smoking status and cervical cancer. (This is the same conclusion that was reached using the 99% confidence interval for the summary odds ratio.)

Exercise 7
a. For individuals who are 35–49 years of age, the estimated odds ratio is

$$\widehat{OR}_1 = \frac{(552)(495)}{(212)(941)}$$
$$= 1.37.$$

For those who are greater than 65 years of age, the estimated odds ratio is

$$\widehat{OR}_2 = \frac{(1,102)(106)}{(87)(1,018)}$$
$$= 1.32.$$

Within each age group, therefore, the odds of suffering from coronary artery disease are greater for individuals with hypertension.
b. To determine whether it is appropriate to combine the information in the two tables, we must conduct a test of homogeneity. The null hypothesis of the test is that the relative odds of CAD are identical in the two age groups. We find that

$$y_1 = \ln(\widehat{OR}_1)$$
$$= \ln(1.37)$$
$$= 0.315,$$

$$y_2 = \ln(\widehat{OR}_2)$$
$$= \ln(1.32)$$
$$= 0.278,$$

$$w_1 = \frac{1}{[(1/552) + (1/212) + (1/941) + (1/495)]}$$
$$= 104.0,$$

and

$$w_2 = \frac{1}{[(1/1102) + (1/87) + (1/1018) + (1/106)]}$$
$$= 43.8.$$

Therefore, the weighted average Y is

$$Y = \frac{\sum_{i=1}^{2} w_i y_i}{\sum_{i=1}^{2} w_i}$$
$$= \frac{(104.0)(0.315) + (43.8)(0.278)}{104.0 + 43.8}$$
$$= 0.304,$$

and the test statistic is

$$X^2 = \sum_{i=1}^{2} w_i (y_i - Y)^2$$
$$= (104.0)(0.315 - 0.304)^2 + (43.8)(0.278 - 0.304)^2$$
$$= 0.04.$$

For a chi-square distribution with 1 df, $p > 0.10$. We do not reject the null hypothesis; therefore, it is appropriate to combine the information in these tables using the Mantel-Haenszel method.

c. The Mantel-Haenszel estimate of the summary odds ratio is

$$\widehat{OR} = \frac{\sum_{i=1}^{2} (a_i d_i / T_i)}{\sum_{i=1}^{2} (b_i c_i / T_i)}$$
$$= \frac{(552)(495)/2200 + (1102)(106)/2313}{(212)(941)/2200 + (87)(1018)/2313}$$
$$= 1.35.$$

d. Since the strata sizes are very large, we may use the normal approximation to construct a confidence interval. Previously we found that $Y = 0.304$. Since

$$\widehat{se}(Y) = \frac{1}{\sqrt{w_1 + w_2}}$$
$$= \frac{1}{\sqrt{104.0 + 43.8}}$$
$$= 0.082,$$

a 95% confidence interval for $\ln(OR)$ is

$$(0.304 - 1.96(0.082), \; 0.304 + 1.96(0.082))$$

or

$$(0.143, \; 0.465).$$

Therefore, a 95% confidence interval for the summary odds ratio itself is

$$(e^{0.143}, \; e^{0.465})$$

or

$$(1.15, \; 1.59).$$

e. To test the null hypothesis that there is no association between hypertension and coronary artery disease, we begin by noting that

$$a_1 \; = \; 552,$$

$$m_1 \; = \; \frac{(1493)(764)}{2200}$$

$$= \; 518.5,$$

$$\sigma_1^{\,2} \; = \; \frac{(1493)(707)(764)(1436)}{(2200)^2(2199)}$$

$$= \; 108.8,$$

$$a_2 \; = \; 1102,$$

$$m_2 \; = \; \frac{(2120)(1189)}{2313}$$

$$= \; 1089.8,$$

and

$$\sigma_2^{\,2} \; = \; \frac{(2120)(193)(1189)(1124)}{(2313)^2(2312)}$$

$$= \; 44.2.$$

The test statistic is

$$X^2 \; = \; \frac{[\sum_{i=1}^{2} a_i - \sum_{i=1}^{2} m_i]^2}{\sum_{i=1}^{2} \sigma_i^2}$$

$$= \; \frac{[(552 + 1102) - (518.5 + 1089.8)]^2}{108.8 + 44.2}$$

$$= \; 13.7.$$

For a chi-square distribution with 1 df, $p < 0.001$. We reject the null hypothesis and conclude that, after adjusting for age, there is an association between hypertension and coronary artery disease. The odds of suffering from CAD are higher for individuals who have hypertension than for those who do not.

Exercise 5

a. A two-way scatter plot of cholesterol versus triglyceride is shown below.

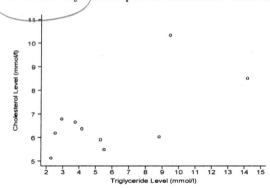

b. There appears to be a slight tendency for cholesterol to increase as triglyceride increases.

c. The mean cholesterol level for these measurements is

$$\bar{x} = \frac{1}{10} \sum_{i=1}^{10} x_i$$
$$= 6.73$$

and the mean triglyceride level is

$$\bar{y} = \frac{1}{10} \sum_{i=1}^{10} y_i$$
$$= 5.91.$$

Since

$$\sum_{i=1}^{10} (x_i - 6.73)(y_i - 5.91) = 34.90,$$

$$\sum_{i=1}^{10} (x_i - 6.73)^2 = 22.00,$$

and

$$\sum_{i=1}^{10} (y_i - 5.91)^2 = 131.22,$$

the Pearson correlation coefficient is

$$r = \frac{\sum_{i=1}^{10}(x_i - 6.73)(y_i - 5.91)}{\sqrt{[\sum_{i=1}^{10}(x_i - 6.73)^2][\sum_{i=1}^{10}(y_I - 5.91)^2]}}$$

$$= \frac{34.90}{\sqrt{(22.00)(131.22)}}$$

$$= 0.650.$$

d. To test the null hypothesis that the population correlation ρ is equal to 0, we calculate the statistic

$$t = r\sqrt{\frac{n-2}{1-r^2}}$$

$$= 0.650\sqrt{\frac{10-2}{1-(0.650)^2}}$$

$$= 2.42.$$

For a t distribution with $10 - 2 = 8$ df, $0.02 < p < 0.05$. Therefore, we reject H_0 and conclude that ρ is not equal to 0; in fact, it is greater than 0.

e. To calculate the Spearman rank correlation, we begin by ordering the two sets of outcomes, ranking them, and calculating the differences in ranks.

Cholesterol	Rank	Triglyceride	Rank	d_i	d_i^2
5.12	1	2.30	1	0	0
5.48	2	5.53	7	−5	25
5.90	3	5.31	6	−3	9
6.02	4	8.83	8	−4	16
6.18	5	2.54	2	3	9
6.36	6	4.18	5	1	1
6.65	7	3.77	4	3	9
6.77	8	2.95	3	5	25
8.51	9	14.20	10	−1	1
10.34	10	9.48	9	1	1

The Spearman rank correlation for these measurements is

$$r_s = 1 - \frac{6\sum_{i=1}^{10} d_i^2}{n(n^2 - 1)}$$

$$= 1 - \frac{6(96)}{10(100 - 1)}$$

$$= 0.418.$$

f. The rank correlation r_s is smaller in magnitude than the Pearson correlation coefficient r. However, it still suggests a moderate positive relationship between cholesterol and triglyceride levels.

g. To test the null hypothesis that ρ is equal to 0 using r_s, we calculate the test statistic

$$t_s = r_s\sqrt{\frac{n-2}{1-r_s^2}}$$

$$= 0.418\sqrt{\frac{10-2}{1-(0.418)^2}}$$

$$= 1.30.$$

In this case, $p > 0.10$ and we are unable to reject the null hypothesis.

Exercise 7

a. Since Apgar score is an ordinal random variable, Spearman's rank correlation must be used. The rank correlation is $r_s = 0.1084$.

```
. spearman sbp apgar5

 Number of obs =      100
Spearman's rho =      0.1084

Test of Ho: sbp and apgar5 independent
      Pr > |t| =      0.2832
```

b. Because the rank correlation is positive, Apgar score tends to increase as systolic blood pressure increases.

c. For the test of the null hypothesis that ρ is equal to 0, $p = 0.2832$. We are unable to reject the null hypothesis.

Exercise 9

a. Between 1991 and 1995, the states with the highest rates of serious actions were Alaska, Oklahoma, West Virginia, Wyoming, and Mississippi, respectively. The states with the lowest rates of serious actions were Rhode Island/South Dakota (tied for 1991), Delaware, District of Columbia, District of Columbia, and Hawaii.

b. The two-way scatter plot is shown below.

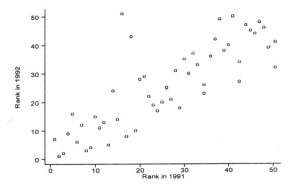

c. There appears to be a strong relationship between the rankings; as 1991 rank increases, 1992 rank increases.

d. The Spearman rank correlation is $r_s = 0.81$.

```
. spearman rank92 rank91

Number of obs =        51
Spearman's rho =       0.8102

Test of Ho: rank92 and rank91 independent
      Pr > |t| =       0.0000
```

e. This correlation is significantly different from 0 ($p < 0.0001$); we conclude that a state's ranking in serious disciplinary actions in 1992 tends to be higher if it also has a higher ranking in 1991.

f. The Spearman rank correlations for 1991 and 1993, 1994, and 1995, respectively, are 0.76, 0.63, and 0.57. The magnitude of the correlation decreases as the years get further apart.

```
. spearman rank93 rank91

Number of obs =        51
Spearman's rho =       0.7643

Test of Ho: rank93 and rank91 independent
      Pr > |t| =       0.0000

. spearman rank94 rank91

Number of obs =        51
Spearman's rho =       0.6292

Test of Ho: rank94 and rank91 independent
      Pr > |t| =       0.0000

. spearman rank95 rank91

Number of obs =        51
Spearman's rho =       0.5736

Test of Ho: rank95 and rank91 independent
      Pr > |t| =       0.0000
```

g. Each of these rank correlations is significantly higher than 0 ($p < 0.0001$).

h. These data suggest that all states are not equally strict in taking disciplinary actions against physicians; some are more strict than others.

Exercise 9

a. The figure below displays the two-way scatter plot of systolic blood pressure versus gestational age. There appears to be a tendency for systolic blood pressure to increase as gestational age increases.

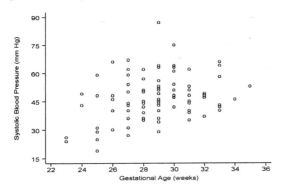

b. The estimated least squares regression line is $\hat{y} = 10.552 + 1.264\,x$.

```
. regress sbp gestage
```

ANOVA Table

Source	SS	df	MS
Model	1016.40959	1	1016.40959
Residual	11856.9504	98	120.98929
Total	12873.36	99	130.033939

Number of obs =	100
F(1, 98) =	8.40
Prob > F =	0.0046
R-squared =	0.0790
Adj R-squared =	0.0696
Root MSE =	11.00

Coefficient Table

sbp	Coef.	Std. Err.	t	P>|t|	[95% Conf. Interval]
gestage	1.26438	.4362311	2.898	0.005	.3986934 2.130066
_cons	10.55207	12.65063	0.834	0.406	-14.55269 35.65682

The slope of the line implies that each one-week increase in gestational age causes an infant's systolic blood pressure to increase by 1.264 mm Hg on average. Although it does not make sense in this case, the y-intercept of 10.552 is the predicted systolic blood pressure for an infant whose gestational age is 0.

c. To test the null hypothesis

$$H_0 : \beta = 0,$$

note that $t = 2.898$ and $p = 0.005$. We reject H_0 at the 0.05 level of significance and conclude that, for low birth weight infants, systolic blood pressure increases in magnitude as gestational age increases.

d. For the population of infants whose gestational age is 31 weeks, the estimated mean systolic blood pressure is

$$\hat{y} = 10.55207 + 1.26438(31)$$
$$= 49.748 \text{ mm Hg.}$$

```
. predict yhat
. list yhat if gestage==31

            yhat
  2.   49.74784
   :
 96.   49.74784
```

e. To construct a 95% confidence interval, we must first find the standard error of the predicted mean $\hat{y}$ when gestational age is 31 weeks.

```
. predict sep, stdp
. list sep if gestage==31

            sep
  2.   1.434265
   :
 96.   1.434265
```

For infants who have a gestational age of 31 weeks, $se(\hat{y}) = 1.434$ mm Hg. Therefore, a 95% confidence interval for the mean value of systolic blood pressure is

$$(49.748 - 1.98(1.434), 49.748 + 1.98(1.434))$$

or

$$(46.91, 52.59).$$

f. For a randomly selected new infant whose gestational age is 31 weeks, the predicted systolic blood pressure is

$$\tilde{y} = 49.748 \text{ mm Hg.}$$

g. To construct a prediction interval, we first find the standard error of $\tilde{y}$.

```
. predict sef, stdf
. list sef if gestage==31

            sef
  2.   11.09263
   :
 96.   11.09263
```

For infants who have a gestational age of 31 weeks, $se(\tilde{y}) = 11.093$ mm Hg. Therefore, a 95% prediction interval for the individual value of systolic blood pressure is

$$(49.748 - 1.98(11.093), 49.748 + 1.98(11.093))$$

or

$$(27.78, 71.71).$$

Note that this interval is much wider than the confidence interval for the mean value of systolic blood pressure.

h. The regression model does not provide an exceptional fit to the observed data. Although gestational age does help to predict systolic blood pressure, only $R^2 = 7.9\%$ of the variability in blood pressure is explained by the linear relationship.

The figure below displays a plot of the residuals versus the fitted values of systolic blood pressure. Although there is one data point that appears to be an outlier, there is no evidence that the assumption of homoscedasticity has been violated or that a transformation of either variable is necessary.

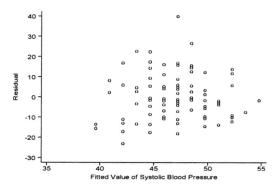

Exercise 11

a. The slope of the line implies that for each 1% increase in contraceptive practice, total fertility rate decreases by 0.062 births per woman. The intercept of 6.83 is the predicted total fertility rate for a country with 0% prevalence of contraceptive practice.

b. The two-way scatter plot is displayed below. It appears that total fertility rate decreases as contraceptive practice increases.

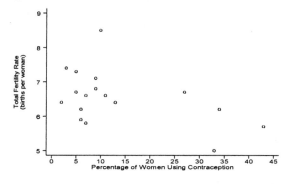

c. The line estimated from countries around the world is shown on the scatter plot on the following page.

d. The total fertility rates for the African nations are higher than would be predicted based on the regression line.

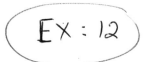

EX : 12

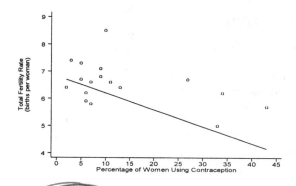

Exercise 13

a. For mean expense per admission, the mean is $2717 and the median is $2600. The minimum and maximum values are $1772 and $4612, respectively. For average length of stay in the hospital, the mean is 7.5 days and the median is 7.7 days. The minimum and maximum values are 5.4 days and 9.7 days.

```
. summarize expadm los, det
```

                      mean expense per admission ($)

	Percentiles	Smallest		
1%	1772	1772		
5%	1974	1859		
10%	2072	1974	Obs	51
25%	2248	2009	Sum of Wgt.	51
50%	2600		Mean	2716.804
		Largest	Std. Dev.	603.9471
75%	3101	3633		
90%	3500	3886	Variance	364752.1
95%	3886	4105	Skewness	.928132
99%	4612	4612	Kurtosis	3.678733

                        average length of stay (days)

	Percentiles	Smallest		
1%	5.4	5.4		
5%	5.7	5.5		
10%	6.3	5.7	Obs	51
25%	6.6	5.9	Sum of Wgt.	51
50%	7.7		Mean	7.490196

		Largest	Std. Dev.	1.015136
75%	8.3	8.7		
90%	8.6	8.9	Variance	1.030502
95%	8.9	9.4	Skewness	-.1632664
99%	9.7	9.7	Kurtosis	2.330016

b. The two-way scatter plot is shown below. The graph suggests that as length of stay increases, average expense per admission increases as well.

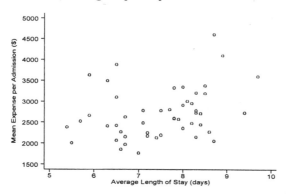

c. The estimated least squares regression line is $\hat{y} = 1281.96 + 191.56\,x$.

. regress expadm los

Source	SS	df	MS		Number of obs =	51
					F(1, 49) =	5.67
Model	1890784.67	1	1890784.67		Prob > F =	0.0212
Residual	16346819.4	49	333608.559		R-squared =	0.1037
					Adj R-squared =	0.0854
Total	18237604.0	50	364752.081		Root MSE =	577.59

expadm	Coef.	Std. Err.	t	P>|t|	[95% Conf. Interval]	
los	191.563	80.4654	2.381	0.021	29.86172	353.2643
_cons	1281.959	608.1041	2.108	0.040	59.92853	2503.99

The slope of the line implies that, for each one-day increase in length of stay, expense per admission increases by \$191.56 on average. The y-intercept of \$1281.96 is the predicted mean value of expense per admission when average length of stay is 0 days.

d. A 95% confidence interval for β is $(29.86, 353.26)$. Note that this interval does not contain the value 0; therefore, expense per admission increases as average length of stay increases.

e. The coefficient of determination is $R^2 = 0.1037$. This coefficient of determination is the square of the Pearson correlation coefficient: $R^2 = r^2$.

f. A plot of the residuals versus the fitted values appears below. A residual plot helps us to determine whether there are any outlying values, whether the assumption of homoscedasticity has been violated, and whether a transformation of variables is necessary.

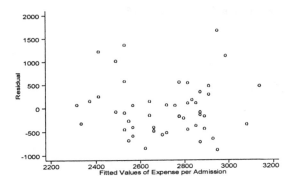

CHAPTER 19

Exercise 7

a. The test statistics and p-values corresponding to each of the four variables are shown below. Because we are not given the sample size, each test statistic is assumed to follow a standard normal distribution.

Variable	Coefficient	Std. Error	t	p
Diet group	−11.25	4.33	−2.60	0.010
Baseline cholesterol	0.85	0.07	12.14	0.000
Body mass index	0.23	0.65	0.35	0.726
Gender	−3.02	4.42	−0.68	0.496

Both diet group and baseline cholesterol level have a significant effect on serum cholesterol level eight weeks after the start of the study.

b. If an individual's body mass index were to increase by 1 kg/m^2, then his or her serum cholesterol level would increase by 0.23 mg/100 ml on average.

c. If an individual's body mass index were to increase by 10 kg/m^2, his or her serum cholesterol level would increase by 2.3 mg/100 ml on average.

d. In this study, a woman is more likely to have a higher serum cholesterol level than a man, all other variables being equal. (Note that the estimated coefficient of gender is negative.) The woman's level would be about 3.02 mg/100 ml higher, on average, although this difference is not statistically significant.

Exercise 9

a. The estimated least squares model is $\hat{y} = 10.007 + 1.263\,x_1 + 1.356\,x_3$.
Since the estimated coefficient of sex is positive, the male infant ($x_3 = 1$) would tend to have the higher systolic blood pressure. Blood pressure would be higher by 1.356 mm Hg on average.

```
. regress sbp gestage sex

    Source |       SS       df       MS                Number of obs =     100
-----------+------------------------------             F(  2,    97) =    4.36
     Model |  1061.73458     2   530.867291            Prob > F      =  0.0154
  Residual |  11811.6254    97   121.769334            R-squared     =  0.0825
-----------+------------------------------             Adj R-squared =  0.0636
     Total |   12873.36     99   130.033939            Root MSE      =  11.035


------------------------------------------------------------------------------
       sbp |      Coef.   Std. Err.       t    P>|t|     [95% Conf. Interval]
-----------+------------------------------------------------------------------
   gestage |   1.262588   .4376449     2.885   0.005     .3939838    2.131192
       sex |   1.356308   2.223097     0.610   0.543    -3.055924     5.76854
     _cons |   10.00706   12.72274     0.787   0.433    -15.24406    35.25818
------------------------------------------------------------------------------
```

b. The two-way scatter plot of systolic blood pressure versus gestational age is displayed below. For males, the fitted least squares line is

$$\hat{y} = 11.363 + 1.263\, x_1.$$

For females, the fitted line is

$$\hat{y} = 10.007 + 1.263\, x_1.$$

The gender difference is not significantly different from 0 ($p = 0.543$).

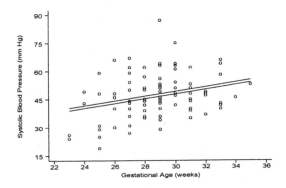

c. When the interaction term is added, the estimated least squares model is
$\hat{y} = 14.981 + 1.090\, x_1 - 15.157\, x_3 + 0.571\, x_1 x_3.$

```
. generate gestsex=gestage*sex
. regress sbp gestage sex gestsex
```

Source	SS	df	MS
Model	1105.44628	3	368.482092
Residual	11767.9137	96	122.582435
Total	12873.36	99	130.033939

```
Number of obs =     100
F(  3,    96) =    3.01
Prob > F      =  0.0341
R-squared     =  0.0859
Adj R-squared =  0.0573
Root MSE      =  11.072
```

sbp	Coef.	Std. Err.	t	P>|t|	[95% Conf. Interval]
gestage	1.090346	.5253655	2.075	0.041	.047504 2.133188
sex	-15.15701	27.74328	-0.546	0.586	-70.22698 39.91296
gestsex	.5714185	.9569067	0.597	0.552	-1.328026 2.470863
_cons	14.98054	15.24191	0.983	0.328	-15.2744 45.23548

The coefficient of the interaction term is not significantly different from 0 ($p = 0.552$); therefore, gestational age does not have a different effect on systolic blood pressure depending on the gender of an infant.

d. The inclusion of the variables **sex** and **gestsex** simultaneously introduces collinearity into the model. Sex and the gestational age–sex interaction term are highly correlated; in fact, their Pearson correlation coefficient is $r = 0.9954$. The addition of the interaction term to the model causes the estimated coefficient of **sex** to change from 1.36 to -15.16 and its standard error to increase by a factor of 12.

Exercise 11

a. A box plot of average salary is displayed below. There appear to be three outlying values, although the rest of the observations are fairly symmetric. The mean and median values are \$14,852 and \$14,573 respectively. The minimum and maximum average salaries are \$11,928 and \$23,594.

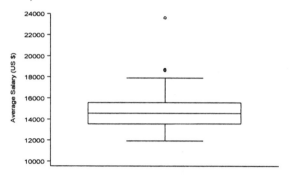

. summarize salary, detail

average salary ($)

	Percentiles	Smallest		
1%	11928	11928		
5%	12737	12508		
10%	12923	12737	Obs	51
25%	13559	12834	Sum of Wgt.	51
50%	14573		Mean	14852.41
		Largest	Std. Dev.	1965.514
75%	15578	17898		
90%	16872	18611	Variance	3863245
95%	18611	18700	Skewness	1.962242
99%	23594	23594	Kurtosis	9.067771

b. The figure on the following page displays the two-way scatter plot of average expense per admission versus salary. The graph suggests that as average salary increases, expense per admission increases as well.

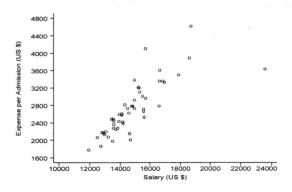

c. The least squares model is $\hat{y} = -2582.74 + 213.80\,x_1 + 0.249\,x_2$.

```
. regress expadm los salary

    Source |       SS       df       MS                Number of obs =      51
-----------+------------------------------             F(  2,    48) =   75.55
     Model | 13840987.8      2  6920493.90             Prob > F      =  0.0000
  Residual | 4396616.24     48  91596.1716             R-squared     =  0.7589
-----------+------------------------------             Adj R-squared =  0.7489
     Total | 18237604.0     50  364752.081             Root MSE      =  302.65


------------------------------------------------------------------------------
    expadm |    Coef.   Std. Err.       t     P>|t|    [95% Conf. Interval]
-----------+------------------------------------------------------------------
       los |  213.7967   42.20769     5.065   0.000    128.9325     298.661
    salary |   .248994   .0217992    11.422   0.000    .2051638    .2928241
     _cons | -2582.736     464.77    -5.557   0.000   -3517.219   -1648.254
------------------------------------------------------------------------------
```

If salary remains constant while length of stay increases by one day, then expense per admission increases by $213.80 on average. Similarly, if length of stay remains constant while average salary increases by $1000, then expense per admission increases by approximately $248.99.
d. When `salary` is added to the model, the coefficient of `los` increases slightly from 191.56 to 213.80.
e. Since the coefficients of `los` and `salary` are both significantly different from 0 at the 0.05 level, the inclusion of salary does improve our ability to predict expense per admission. Furthermore, R^2 increases from 10.4% to 75.9%!

f. A plot of the residuals versus the fitted values of expense per admission is shown below.

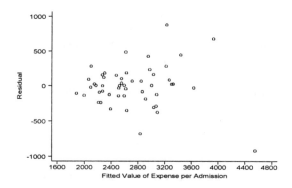

There does appear to be one outlying value (as was also evident in the original scatter plot). Ignoring this observation, however, the assumption of homoscedasticity has not been violated, and a transformation of variables is not necessary.

CHAPTER 20

Exercise 5

a. As the number of prenatal visits to the doctor increases, the probability that a child will be born with congenital syphilis decreases. (Note that the estimated coefficient of number of visits is negative.)

b. The relative odds that a newborn will suffer from syphilis for unmarried versus married mothers is

$$e^{0.779} = 2.18.$$

It appears that the children of unmarried mothers are more likely to be born with syphilis. However, we would need the standard error of this estimated coefficient to determine whether the effect is statistically significant.

c. If a woman uses either cocaine or crack, the relative odds that her child will be born with congenital syphilis is

$$e^{1.354} = 3.87.$$

d. A 95% confidence interval for the coefficient itself is

$$(1.354 - 1.96(0.162), 1.354 + 1.96(0.162))$$

or

$$(1.036, 1.672).$$

Therefore, a 95% confidence interval for the true population odds ratio that a child will be born with syphilis for women who have used cocaine or crack versus those who have not is

$$(e^{1.036}, e^{1.672})$$

or

$$(2.82, 5.32).$$

Exercise 7

a. The fitted logistic regression model is

$$\ln\left[\frac{\hat{p}}{1 - \hat{p}}\right] = -0.3037 - 0.2496\, x_1.$$

For each one unit increase in apgar score, the log odds of experiencing a germinal matrix hemorrhage decrease by 0.2496, on average.

```
. logit grmhem apgar5

Iteration 0:   log likelihood = -42.270909
Iteration 1:   log likelihood = -39.727053
Iteration 2:   log likelihood = -39.463638
Iteration 3:   log likelihood = -39.463411
```

```
Logit estimates                               Number of obs   =        100
                                              LR chi2(1)      =       5.61
                                              Prob > chi2     =     0.0178
Log likelihood = -39.463411                   Pseudo R2       =     0.0664
```

```
------------------------------------------------------------------------
 grmhem |    Coef.    Std. Err.      z     P>|z|     [95% Conf. Interval]
--------+---------------------------------------------------------------
 apgar5 | -.2496074   .1043609    -2.392   0.017    -.4541511   -.0450638
  _cons | -.3037311   .619057     -0.491   0.624    -1.51706     .9095984
------------------------------------------------------------------------
```

b. For a child with a five-minute apgar score of 3,

$$\ln\left[\frac{\hat{p}}{1-\hat{p}}\right] = -0.3037 - 0.2496(3)$$

$$= -1.0525.$$

Therefore,

$$\frac{\hat{p}}{1-\hat{p}} = e^{-1.0525}$$

$$= 0.3491,$$

and the estimated probability that the child will experience a hemorrhage is

$$\hat{p} = \frac{0.3491}{1.3491}$$

$$= 0.259.$$

For a child with an apgar score of 7,

$$\ln\left[\frac{\hat{p}}{1-\hat{p}}\right] = -0.3037 - 0.2496(7)$$

$$= -2.0509.$$

Therefore,

$$\frac{\hat{p}}{1-\hat{p}} = e^{-2.0509}$$

$$= 0.1286,$$

and the estimated probability that the child will experience a hemorrhage is

$$\hat{p} = \frac{0.1286}{1.1286}$$

$$= 0.114.$$

c. To test the null hypothesis

$$H_0 : \beta_1 = 0,$$

85

note that $z = -2.392$ and $p = 0.017$. We reject the null hypothesis at the 0.05 level and conclude that β_1 is not equal to 0. As five-minute apgar score increases, the log odds of experiencing a germinal matrix hemorrhage decrease.

d. The fitted model is

$$\ln\left[\frac{\hat{p}}{1-\hat{p}}\right] = -1.5353 - 1.4604\, x_2.$$

```
. logit grmhem tox

Iteration 0:   log likelihood = -42.270909
Iteration 1:   log likelihood = -41.040989
Iteration 2:   log likelihood = -40.927887
Iteration 3:   log likelihood = -40.924735
Iteration 4:   log likelihood = -40.924731


Logit estimates                                 Number of obs   =        100
                                                LR chi2(1)      =       2.69
                                                Prob > chi2     =     0.1008
Log likelihood = -40.924731                     Pseudo R2       =     0.0318


------------------------------------------------------------------------------
  grmhem |      Coef.   Std. Err.       z    P>|z|     [95% Conf. Interval]
---------+--------------------------------------------------------------------
     tox | -1.460402   1.066213    -1.370   0.171    -3.550141    .6293363
   _cons |  -1.53533    .2946408    -5.211   0.000    -2.112815   -.9578446
------------------------------------------------------------------------------
```

The log odds of experiencing a germinal matrix hemorrhage are lower by 1.4604, on average, for a child whose mother experienced toxemia during pregnancy.

e. For a child whose mother was diagnosed with toxemia during pregnancy,

$$\ln\left[\frac{\hat{p}}{1-\hat{p}}\right] = -1.5353 - 1.4604(1)$$
$$= -2.9957.$$

Therefore,

$$\frac{\hat{p}}{1-\hat{p}} = e^{-2.9957}$$
$$= 0.0500,$$

and the estimated probability that the child will experience a germinal matrix hemorrhage is

$$\hat{p} = \frac{0.0500}{1.0500}$$
$$= 0.048.$$

For a child whose mother was not diagnosed with toxemia,

$$\ln\left[\frac{\hat{p}}{1-\hat{p}}\right] = -1.5353 - 1.4604(0)$$
$$= -1.5353.$$

Therefore,

$$\frac{\hat{p}}{1-\hat{p}} = e^{-1.5353}$$

$$= 0.2154,$$

and the estimated probability that the child will experience a germinal matrix hemorrhage is

$$\hat{p} = \frac{0.2154}{1.2154}$$

$$= 0.177.$$

f. The estimated odds of suffering a germinal matrix hemorrhage for infants whose mothers were diagnosed with toxemia relative to those whose mothers were not are

$$\widehat{OR} = e^{-1.4604}$$

$$= 0.232.$$

g. A 95% confidence interval for β_2 is

$$(-1.4604 - 1.96(1.0662)\,,\, -1.4604 + 1.96(1.0662))$$

or

$$(-3.5502\,,\, 0.6294).$$

Therefore, a 95% confidence interval for the population relative odds itself is

$$(e^{-3.5502}\,,\, e^{0.6294})$$

or

$$(0.029\,,\, 1.876).$$

Note that the interval does contain the value 1; this indicates that toxemia does not have a statistically significant effect on the probability of experiencing a germinal matrix hemorrhage.

Exercise 9

a. The three logistic regression models are shown below. The log odds of experiencing peritonitis are higher for females versus males and for non-whites versus whites, and are lower for older individuals.

```
. logit perito age

Iteration 0:   log likelihood = -29.720163
Iteration 1:   log likelihood = -29.714509
Iteration 2:   log likelihood = -29.714509

Logit estimates                          Number of obs   =        46
                                         LR chi2(1)      =      0.01
                                         Prob > chi2     =    0.9153
Log likelihood = -29.714509              Pseudo R2       =    0.0002
```

```
-------------------------------------------------------------------------------
  perito |     Coef.    Std. Err.       z      P>|z|      [95% Conf. Interval]
---------+---------------------------------------------------------------------
     age | -.0027128    .0255417    -0.106    0.915     -.0527736     .047348
   _cons |  .7704915    1.372803     0.561    0.575     -1.920154    3.461137
-------------------------------------------------------------------------------

. logit perito sex

Iteration 0:   log likelihood = -29.720163
Iteration 1:   log likelihood = -28.963006
Iteration 2:   log likelihood = -28.959612
Iteration 3:   log likelihood = -28.959612

Logit estimates                              Number of obs   =          46
                                             LR chi2(1)      =        1.52
                                             Prob > chi2     =      0.2175
Log likelihood = -28.959612                  Pseudo R2       =      0.0256


-------------------------------------------------------------------------------
  perito |     Coef.    Std. Err.       z      P>|z|      [95% Conf. Interval]
---------+---------------------------------------------------------------------
     sex |  .7884574    .6513389     1.211    0.226     -.4881435    2.065058
   _cons |  .3101549    .3969581     0.781    0.435     -.4678687    1.088179
-------------------------------------------------------------------------------

. logit perito race

Iteration 0:   log likelihood = -29.720163
Iteration 1:   log likelihood = -28.788539
Iteration 2:   log likelihood = -28.770971
Iteration 3:   log likelihood = -28.770941

Logit estimates                              Number of obs   =          46
                                             LR chi2(1)      =        1.90
                                             Prob > chi2     =      0.1683
Log likelihood = -28.770941                  Pseudo R2       =      0.0319


-------------------------------------------------------------------------------
  perito |     Coef.    Std. Err.       z      P>|z|      [95% Conf. Interval]
---------+---------------------------------------------------------------------
    race |  1.098612    .8544906     1.286    0.199     -.5761585    2.773383
   _cons |  .4054651    .3450328     1.175    0.240     -.2707867    1.081717
-------------------------------------------------------------------------------
```

b. For a white patient undergoing dialysis,

$$\ln\left[\frac{\hat{p}}{1-\hat{p}}\right] = 0.4055 + 1.0986(0)$$

$$= 0.4055.$$

Therefore,

$$\frac{\hat{p}}{1-\hat{p}} = e^{0.4055}$$

$$= 1.500,$$

and the estimated probability that the patient experiences peritonitis is

$$\hat{p} = \frac{1.500}{2.500}$$

$$= 0.600.$$

For a non-white patient,

$$\ln\left[\frac{\hat{p}}{1-\hat{p}}\right] = 0.4055 + 1.0986(1)$$

$$= 1.5041.$$

Therefore,

$$\frac{\hat{p}}{1-\hat{p}} = e^{1.5041}$$

$$= 4.500,$$

and the estimated probability that the patient experiences peritonitis is

$$\hat{p} = \frac{4.500}{5.500}$$

$$= 0.818.$$

c. The odds of developing peritonitis for females versus males are

$$e^{0.788} = 2.20.$$

d. At the 0.05 level of significance, none of the three variables helps to predict peritonitis in patients undergoing dialysis.

e. Not all patients are followed for the same length of time (minimum follow-up is two years, but there is no maximum). It is possible that patients who are followed for longer periods of time are more likely to experience peritonitis than those followed for shorter periods.

CHAPTER 21

Exercise 7

a. Five deaths were observed in this sample of ten patients, and five observations were censored.

b. The product-limit estimate of the survival function $S(t)$ is presented in the following table.

Time	q_t	$1 - q_t$	$S(t)$
0.0	0.0000	1.0000	1.0000
0.5	0.0000	1.0000	1.0000
1.0	0.3333	0.6667	0.6667
2.0	0.1667	0.8333	0.5556
5.0	0.0000	1.0000	0.5556
8.0	0.0000	1.0000	0.5556
9.0	0.3333	0.6667	0.3704
10.0	0.0000	1.0000	0.3704
12.0	0.0000	1.0000	0.3704

c. The estimated probability of survival at 1 month is $\widehat{S}(1) = 0.6667$. The estimated probability of survival at 5 months is $\widehat{S}(5) = 0.5556$, and at 6 months is $\widehat{S}(6) = 0.5556$.

d. The survival curve is displayed below.

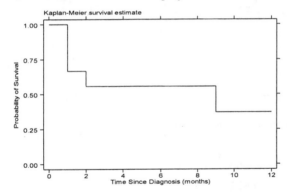

Exercise 9

a. Kaplan-Meier estimates of the two survival curves for time to recurrence of first tumor are shown on the following page.

. sts list if group==1

```
        failure _d:  censor
  analysis time _t:  time
```

Time	Beg. Total	Fail	Net Lost	Survivor Function	Std. Error	[95% Conf. Int.]	
1	47	1	1	0.9787	0.0210	0.8584	0.9970
2	45	4	0	0.8917	0.0457	0.7592	0.9535
3	41	7	0	0.7395	0.0647	0.5870	0.8428
4	34	0	1	0.7395	0.0647	0.5870	0.8428
5	33	2	0	0.6947	0.0681	0.5394	0.8065
6	31	2	0	0.6498	0.0707	0.4932	0.7689
7	29	1	1	0.6274	0.0717	0.4706	0.7496
9	27	2	0	0.5810	0.0735	0.4242	0.7090
10	25	1	1	0.5577	0.0742	0.4015	0.6882
12	23	2	0	0.5092	0.0752	0.3547	0.6444
14	21	0	1	0.5092	0.0752	0.3547	0.6444
16	20	1	0	0.4838	0.0757	0.3304	0.6212
17	19	1	0	0.4583	0.0758	0.3065	0.5976
18	18	1	1	0.4328	0.0758	0.2832	0.5736
23	16	0	1	0.4328	0.0758	0.2832	0.5736
25	15	1	0	0.4040	0.0760	0.2562	0.5470
26	14	0	1	0.4040	0.0760	0.2562	0.5470
28	13	1	0	0.3729	0.0763	0.2274	0.5184
29	12	1	3	0.3418	0.0760	0.1997	0.4890
32	8	0	1	0.3418	0.0760	0.1997	0.4890
34	7	0	1	0.3418	0.0760	0.1997	0.4890
35	6	1	0	0.2849	0.0819	0.1398	0.4486
36	5	0	1	0.2849	0.0819	0.1398	0.4486
37	4	0	1	0.2849	0.0819	0.1398	0.4486
41	3	0	1	0.2849	0.0819	0.1398	0.4486
49	2	0	1	0.2849	0.0819	0.1398	0.4486
59	1	0	1	0.2849	0.0819	0.1398	0.4486

. sts list if group==2

```
        failure _d:  censor
  analysis time _t:  time
```

Time	Beg. Total	Fail	Net Lost	Survivor Function	Std. Error	[95% Conf. Int.]	
1	38	2	2	0.9474	0.0362	0.8056	0.9866
2	34	4	0	0.8359	0.0613	0.6704	0.9228
3	30	1	0	0.8080	0.0653	0.6388	0.9036
4	29	2	0	0.7523	0.0717	0.5777	0.8628

5	27	1	0	0.7245	0.0743	0.5481	0.8413
6	26	2	0	0.6687	0.0783	0.4907	0.7966
9	24	0	1	0.6687	0.0783	0.4907	0.7966
10	23	0	1	0.6687	0.0783	0.4907	0.7966
13	22	0	1	0.6687	0.0783	0.4907	0.7966
17	21	2	0	0.6050	0.0828	0.4245	0.7448
18	19	0	1	0.6050	0.0828	0.4245	0.7448
22	18	1	1	0.5714	0.0848	0.3903	0.7169
24	16	1	0	0.5357	0.0867	0.3544	0.6869
25	15	0	3	0.5357	0.0867	0.3544	0.6869
26	12	1	0	0.4911	0.0902	0.3073	0.6514
38	11	1	1	0.4464	0.0924	0.2635	0.6140
41	9	0	2	0.4464	0.0924	0.2635	0.6140
44	7	0	1	0.4464	0.0924	0.2635	0.6140
45	6	0	1	0.4464	0.0924	0.2635	0.6140
46	5	0	1	0.4464	0.0924	0.2635	0.6140
49	4	0	1	0.4464	0.0924	0.2635	0.6140
50	3	0	1	0.4464	0.0924	0.2635	0.6140
54	2	0	1	0.4464	0.0924	0.2635	0.6140
59	1	0	1	0.4464	0.0924	0.2635	0.6140

----

b. Survival curves for each of the two groups of patients are shown below.

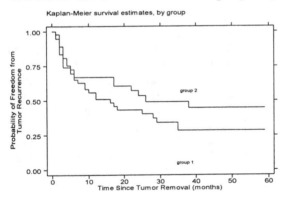

c. Based on the curves, patients treated with the drug might have somewhat longer survival times.

d. We use the log-rank test to test the null hypothesis

$$H_0 : S_1(t) = S_2(t).$$

Since $p = 0.2175$, we are unable to reject the null hypothesis. These data do not provide evidence that the distributions of recurrence times differ by treatment group.

```
. sts test group, logrank

        failure _d:  censor
   analysis time _t:  time


Log-rank test for equality of survivor functions
--------------------------------------------------

        |  Events
 group  |  observed      expected
--------+---------------------------
 1      |     29          24.91
 2      |     18          22.09
--------+---------------------------
 Total  |     47          47.00

            chi2(1) =      1.52
            Pr>chi2 =     0.2175
```

e. We again use the log-rank test.

Since $p = 0.6816$, we are unable to reject the null hypothesis. Within the placebo group, the data do not provide evidence that the distributions of recurrence times differ by number of tumors.

```
. sts test number if group==1, logrank

        failure _d:  censor
   analysis time _t:  time


Log-rank test for equality of survivor functions
--------------------------------------------------

        |  Events
 number |  observed      expected
--------+---------------------------
 1      |     16          14.93
 2      |     13          14.07
--------+---------------------------
 Total  |     29          29.00

            chi2(1) =      0.17
            Pr>chi2 =     0.6816
```

CHAPTER 22

Exercise 5

a. Nonresponse might bias the results of the study. It is conceivable, for example, that mothers who use either marijuana or cocaine might be more likely to refuse to respond to questions about drug use; in this case, an estimate of drug use prevalence based on the survey would underestimate the true population prevalence.

b. Interviewing only those mothers who agree to be questioned is unlikely to provide a representative sample from the population of expectant mothers. Again, we might expect that women who agree to be interviewed are less likely to use either marijuana or cocaine, while those who refuse are more likely to use drugs.

Solutions are not provided for the remaining exercises, which do not have a single correct answer.